D1351274

# FROZEN IN TIME

Ian McCaskill and Paul Hudson

GREAT NORTHERN

# FROZEN IN TIME

Ian McCaskill and Paul Hudson

Great Northern Books
PO Box 213, Ilkley, LS29 9WS
www.greatnorthernbooks.co.uk

First published 2006



ISBN: 1 905080 09 3

Design and layout: David Burrill

Printed by Quebecor Ibérica, Barcelona

CIP Data
A catalogue for this book is available from the British Library

# CONTENTS

# Prologue

We all have memories of sledging, snowball fights, building snowmen and sucking icicles. It's a part of everyone's childhood - or at least it used to be - winters have been so mild in recent years that many British children have scarcely experienced snow at all.

My memories of winters start when I was eight years old with one of the worst of them all - 1947. I was already interested in the weather and keeping a record of daily temperatures and I can remember the shock when I discovered that the temperature even inside our Glasgow house was well below freezing and the snowfall was heavy and prolonged. We didn't have telly weathermen in those days; we didn't even have telly. But looking back, it was clear that the snow must have arrived from the moist and still relatively warm Atlantic, crashing into cold continental air sitting over Scotland. The reports in the press and on the radio suggested that England and Wales were suffering the same weather. The result in Scotland was extensive drifting snow with banks of three feet or more piled up at the edge of the pavement. When you are just eight years old and only three feet tall, it leaves a mark!

My six year-old brother and I had another good reason to remember that winter. This was the era when children were seen but emphatically not heard, and were told nothing and knew less, so when my mother wrapped us in our warmest clothes, marched us through the snowy, frozen streets and put us on the tram in south Glasgow we were not told the reason for our journey. The tram rattled slowly through the snow ten miles to Renfrew, where our dear Aunt Ella was waiting to meet us. With a hug and a kiss, but without a word, Mother jumped straight back on the tram and disappeared. Our self control vanished as we saw the tram receding from us, outlined against the snow. Despite all Aunt Ella's efforts to console us, we clung to each other, sobbing and howling, not knowing why we had been exiled from our home. Mother left us in Renfrew for two weeks of dumplings and rabbit stew - the only meat available that was not rationed - but when we were eventually summoned back to Glasgow, we at last discovered the reason for our absence when we were greeted by a much slimmer mum nursing our brand new baby sister.

The winter of 1963 was very different in some ways, though every bit as severe and prolonged. An area of high pressure usually forms in winter over Siberia beneath a blanket of stagnant and bitterly cold air in a classic "chicken and egg" meteorological situation, but on occasions - and 1963 was one of them - the area is even colder and much larger than usual, spreading right across the North Sea to cover the British Isles. On the southern flank of the Siberian high there is a wide belt of cold easterly winds that are just a smidgin less cold than over Central Europe when they reach Britain, but are much moister, producing, in my view, much worse winter weather - less snow than in many winters, but dull, dreary and raw, with flurries of snow.

This situation can last for weeks on end... and did. At the time I was a forecaster at RAF Leuchars - I was in shock after coming there from my previous posting to Cyprus. Leuchars was an important base at the time - a "master diversion" airfield, manned by thousands of personnel, that held a vital part of our Cold War defence against possible nuclear attack. However, the phrase "Cold War" took on an entirely new meaning that winter. Leuchars was a forward outpost of fighter aircraft, poised 24/7 to intercept anything hostile approaching from the East. But morning briefings were sad, short affairs that winter, as the prolonged sub-zero temperatures and falls of snow made scrambling anything other than eggs impossible. We even tried taxiing the fighters slowly up and down the runways, gunning their engines, while the jet-wash blasted the surface. The ice melted, but only briefly, and then re-froze into a perfect, 1,000 foot ice-rink. Nothing the wit or muscle of man could devise could free those runways from ice and snow. As fast as one part was cleared, another would freeze and, in a winter where even the sea froze for a hundred yards offshore, using salt to melt the ice proved as ineffective as jet-wash. Nothing at all landed or took off from those runways for six weeks. Had the Soviets chosen to attack during that period, we couldn't have done a thing about it.

During the 1978-79 "Winter of Discontent" I was living on the fringes of the Peak District in Derbyshire. I can remember the snowdrifts lying for weeks on end - some of the North-facing gullies still had snow in them in June, but it wasn't the snow, cold and ice, nor the frozen pipes and blocked roads, nor the endless series of strikes by tanker drivers, road hauliers, hospital workers, dustbin men, gravediggers and, it seemed, just about everybody else, that made that winter so bitter for me. I could have survived all that without complaint, but when the weather caused the local pubs to close because their pipes were frozen and the brewery drays couldn't get through the snow, it was time to give emigration to warmer climes some serious thought!

**Ian McCaskill**

# PART I

# The Worst Winters in History

It's the old codger's lament: Things aren't what they used to be - 'We don't get summers/pork pies/films/footballers/music/cups of tea like we did when I was young'. In most cases, it's just the rose-tinted nostalgia of the old for the lost years of their youth, but in one case it is actually true: we really don't get winters like we used to any more.

Savage winters have been a recurring feature of our history as long as records have been kept. The Thames was frozen for nine weeks in 250 AD, and for thirteen weeks in 923 AD, when carts and wagons crossed the ice. In 1410, the river was frozen for fourteen weeks and winters were even more severe during the "Little Ice Age" between the middle of the sixteenth century and the middle of the nineteenth century. During the winter of 1536, Henry VIII was said to have travelled in a sleigh along the frozen River Thames all the way from the centre of London to Greenwich, and Elizabeth I took walks on the ice during the winter of 1564. The early years of the 1660s saw a succession of harsh winters. Skating was said to have been introduced into England during the winter of 1662/63; King Charles II watched the new sport on the frozen Thames. In the bitter winter of 1664-65, a severe frost lasting about two months included what was reputedly 'the coldest day ever in England'. The Thames froze from bank to bank above old London Bridge and a Frost Fair was held on ice 'so thick as to bear not only streets of booths in which they roasted meat, and divers shops of wares quite as in a town, but coaches, carts and horses.' Carriages took shortcuts across the frozen river rather than braving the traffic jams of vehicles - carts, carriages and wagons - and the droves of animals and hordes of pedestrians that, then as now, reduced traffic over the bridge to a crawl. There was even a bonfire on the ice, not perhaps the wisest place to light one, but it passed off without incident.

The winter of 1683-84 may have been the coldest ever. The 'Great Frost' began in mid-December 1683 and by early January the Thames was frozen all the way to London Bridge. The diarist John Evelyn called the winter 'intolerably severe'. The frost, claimed to be the longest on record, kept the Thames frozen for two months, with as much as eleven inches of solid ice covering the surface. Evelyn rode in a coach along the frozen river all the way from London Bridge to Lambeth, 'Coaches plied from Westminster to the Temple, and from several other stairs too and fro, as in the streets, sleds, sliding with skates, bull-baiting, horse and coach races, puppet plays and interludes, cooks, tippling and other lewd places, so that it

*Possibly the coldest winter ever - the Thames during "The Great Frost of 1739". In the foreground a Frost Fair spreads across the ice and to the right figures can be seen inspecting the piers of the still incomplete Westminster Bridge. St Paul's dominates the London skyline in the background.*

seemed to be a bacchanalian triumph, or carnival on the water.' It was so cold that by January the Company of Watermen was distributing £200 to 'poor watermen' deprived by the ice of their normal source of income - the river.

Near Manchester, the ground froze to a depth of twenty-seven inches, and in Somerset, to an incredible depth of four feet. RD Blackmore, in Lorna Doone, described the winter on Exmoor. "An ancient shepherd had dropped in, and taken supper with us, and foretold a heavy fall [of snow] and great disaster to livestock. He said that he had known a frost beginning, just as this had done, with a black east wind, after days of raw cold fog, and then on the third night of the frost at this very time of year (to wit the 15th of December) such a snow set in as killed half of the sheep, and many even of the red deer and the forest ponies... Two of his toes had been lost by frost-nip, while he dug out his sheep...

"All the birds were set in one direction, steadily journeying westward, not with any heat

*The last ever great frost fair held on the Thames in 1814. A fire is blazing in the tent in the right foreground, to the left are street vendors and a printing press set up on the ice. Ships are frozen into the river but the ice is thin enough for people to be depicted falling through it. In the background is Blackfriars Bridge.*

of speed, neither flying far at once; but all (as if on business bound) partly running, partly flying, partly fluttering along; silently, and without a voice, neither pricking head nor tail. This movement of the birds went on, even for a week or more; every kind of thrushes passed us, every kind of wild fowl, even plovers went away, and crows and snipes and woodcocks. And before half the frost was over, all we had in the snowy ditches were hares so tame that we could pat them; partridges that came to hand, with a dry noise in their crops; heath-poults, making cups of snow; and a few poor hopping redwings, flapping in and out of the hedge, having lost the power to fly. And all the time, their great black eyes, set with gold around them, seemed to look at any man for mercy and for comfort...

"Of the sheep upon the mountain, and the sheep upon the western farm, and the cattle on the upper burrows, scarcely one in ten was saved... The great snow never ceased a moment for three days and nights, and then when all the earth was filled, and topmost hedges were unseen, and the trees broke down with weight (wherever the wind had not lightened them), a brilliant sun broke forth and showed the loss of all our customs. All our house was quite snowed up, except where we had purged a way by dint of constant shovellings, the kitchen was as dark and darker than the cider-cellar, and long lines of furrowed scollops ran even up to the chimney stacks. Several windows fell right inwards, through the weight of the snow against them; and the few that stood bulged in and bent like an old bruised lanthorn [lantern]. We were obliged to cook by candlelight; we were forced to read by candlelight; as for baking, we could not do it, because the oven was too chill.

"That night such a frost ensued as we had never dreamed of, neither read in ancient books, or histories of [the Arctic voyages of] Frobisher. The kettle by the fire froze, and the crock upon the hearth-cheeks; many men were killed, and cattle rigid in their head-ropes. Then I heard that fearful sound, which never I heard before, neither since have heard (except during that same winter), the sharp yet solemn sound of trees, burst open by the frost-blow. Our great walnut lost three branches, and has been dying ever since... and the ancient oak at the cross was rent, and many score of ash-trees... It mattered not what way the wind was. Often and often the vanes went around, and we hoped for change of weather: the only change was that it seemed (if possible) to grow colder... But why should I tell all this? The people who have not seen it (as I have) will only make faces and disbelieve, till such another frost comes; which perhaps may never be."

The winter of 1694-95 saw five weeks of 'continual snows', and ended only around mid-April, by which time Britain had endured almost six months of winter, while the unfortunate inhabitants of Iceland had seen the Arctic sea ice extend far south of its usual limit, to encircle the entire Icelandic coast. Six winters in that decade were very severe and the cumulative effect of the run of poor harvests saw many peasant farmers starve to death.

In 1715-16 the Thames was frozen for two months, and a frost fair took place on the ice. On 25 January the ice was 'lifted by some fourteen feet by a flood tide' but it was so thick that it did not break. 1739-40 saw one of the worst - and certainly one of the coldest - winters since records began. Winter came unusually early - on 8 October - and at one point the Thames remained frozen for eight weeks. A violent easterly gale on 29-30 December drove large blocks of drifting ice before it and caused substantial damage to the piers of London Bridge and shipping on the Thames. Many ships were driven ashore and dashed to pieces. 16 January 1740 was described as 'the coldest day in the memory of man', Spring was also cold and miserable and the following October was the coldest on record, with ice already forming on rivers. After two successive years of terrible winters and failed harvests, 1741 was known in Ireland as bliadhain an áir - the year of the slaughter - with an estimated 300,000 dead of starvation, ten percent of the total population. A pamphlet, "The Groans of Ireland", described 'Want and misery in every face... the roads spread with dead and dying bodies; mankind the colour of the docks and nettles which they fed on; two or three, sometimes more, in a cart going to the grave for want of bearers to carry them, and many buried only in the fields and ditches where they perished.'

In 1784-86, two successive severe winters followed the repeated eruptions of the Icelandic volcano Laki, in 1783 and 1784. In Iceland itself, the winter of 1783-84 was known as "The Famine Winter" in which a quarter of the population died. Reports from all over Europe spoke of 'a blue haze' or 'dry-fog' in the atmosphere and damage to plants and crops, and there were numerous complaints of respiratory problems. The effect on the British weather was equally pronounced. The Thames froze completely in both winters, there were near-continuous frosts, and lying snow for four months. Sleet was recorded near the coast of the Moray Firth in August 1784 and heavy snow fell in the South on 7 October. The last substantial snowfall of the winter came six months later on 4 April.

In January 1789, during another Frost Fair on the Thames, melting ice dragged a ship anchored to a riverside public house, pulling the building down and crushing five people to death. 1794-95 was another exceptionally severe winter, beginning on Christmas Eve and lasting until late March. It included the coldest January since weather recording by instruments began in 1659. The Severn and The Thames froze, and Frost Fairs were held.

In 1813-1814, thick fog and sharp frosts on 27 December heralded the start of a bitterly cold winter. The fog was so dense that the Birmingham mail coach took seven hours to travel the sixteen miles from Central London to Uxbridge in Middlesex. On 30 and 31 December, traffic was almost at a standstill in London and coachmen had to dismount and lead their horses through the streets. Bitter northerly winds cleared the fog in early January, but brought heavy snowfall and frosts so severe that ice floating down the Thames during a brief thaw between 5 and 7 February caused severe damage to shipping. The tidal stretch of the Thames froze and the last great Frost Fair was held on the river. An elephant was led across the river below Blackfriars Bridge, and a printer named Davis published a book entitled Frostiana; Or a History of the River Thames in a Frozen State. The narrow arches and thick stone piers of the old London Bridge had greatly restricted the river flow, making it far more likely to freeze. The increased flow after the construction of the the new London Bridge prevented ice from forming to anything like the same extent and no more great Frost Fairs were ever held.

The following year, the eruption of Tambora in the East Indies on 10 April 1815 killed 90,000 people and blasted debris and millions of tons of sulphur dioxide into the stratosphere. While the fine particles suspended in the air produced some spectacular sunsets worldwide for over a year afterwards, the eruption also disrupted global weather patterns and made Britain unusually cold and wet. The following year, 1816, was known as "the year of no summer" and was followed by a bleak and very prolonged winter. There was ice on the Thames on 2 September of that year and on 20 October 'a great hurricane and snowstorm' struck northern Scotland. 'The stooks of corn were yet out in the fields, and the snow had to be cast to get at them; when dug out they were a frozen lump, and could not be thawed for the cattle.' There were still snow drifts on Helvellyn in the Lake District on 30 July of the

*A caricature of Patrick Murphy, "the Dick-tater, alias the weather-cock of the walk", the celebrated prophet of "Murphy's Winter" of 1837-38. Master of all weathers, a weather vane sticks out of the top of his potato head, he has a globe for a body, and holds bellows and a telescope under his right arm, and an umbrella and a watering can under his left. His right foot is bare, and his left foot is in a snow shoe.*

following year. The harvest across Europe was delayed and in Ireland and western parts of Britain there was a total crop failure, leading to widespread starvation.

1837-38, "Murphy's Winter", was so named in honour of Patrick Murphy, who won fame and fortune from the sale of an almanac in which he predicted the severe frost of January 1838. 20 January saw temperatures as low as -16°C in London, the coldest recorded in the capital throughout the 19th century, and even at midday the temperature at Greenwich had risen no higher than -11°C. Spring planting was delayed, and frosts in August, continuing through September and October wrecked the harvest and caused frost-damage to root crops leading to severe hardship and even starvation for tenant farmers.

In 1875-76, significant snowfall was recorded in November, December, January, February, March, April and May, making it one of the most prolonged winters ever. In the first week of December blizzards left one to two feet of level snow over the South-East and as late as April, snowstorms dumped almost two feet of snow on the Midlands.

*An 1870s view of the frozen River Thames, looking north from just below London Bridge.*

In December 1894 'the worst storm of the century' struck Britain, toppling thousands of trees and damaging roofs and church spires throughout the country. The calm that followed heralded the start of 'the twelve week frost' beginning on 30 December. In mid-February the whole of the Thames was said to be blocked by ice-floes six or seven feet thick. An ox was roasted on the river at Kingston and coaches crossed on the ice at Oxford. Thousands skated on the frozen Serpentine in London and on Windermere in the Lake District, which was frozen solid for six weeks. Thousands of people arrived in the Lake District by train, bands played in Bowness Bay and crowds of over two thousand people skated on the ice, rode bicycles fitted with spiked tyres, sailed 'ice-yachts', and ate and drank at stalls erected on the frozen lake.

In the twentieth century, 1916-17 was the year in which thousands of soldiers already enduring the miseries of trench warfare and the after math of the disastrous Battle of the Somme on the Western Front, had to suffer 'the worst winter in twenty years' as well. Soldiers 'lying at each listening post were freezing stiff, and would take half an hour's buffeting and rubbing on return to avoid becoming casualties.' Many men on both sides died

*The frozen river Thames in 1895.*

*A train on the Highland Railway in Scotland, near Altnabreac in 1895, where the drifts were forty to fifty feet deep. Drivers of steam engines fitted with snowploughs - first used in the 1850s - often tried to ram their way through drifts but some were derailed in the process.*

*A man standing in a rowing boat frozen into the river, surveys the towering ice-floes on the Thames in 1895.*

of frostbite and hypothermia, and in conditions where even survival was a daily battle, both sides restricted their aggressive actions to sporadic sniping, artillery duels and small raids on the enemy trenches. An army chaplain, David Railton, described the conditions in a letter to his wife. 'Today there was an inspection. The men were not kept there long. Still two of them fainted from cold.... Even the Sacred Wine in the Chalice froze on Sunday. Our men are very, very brave. Some of them could not bear their boots on this last time up the line. They put sandbags around their feet... We have thirty degrees of frost out here.'

A soldier in the reserve areas 'had to light a fire of paper in boots frozen hard before they

*A boy walking his dog though the deep snow covering the beach at Whitby in the winter of 1904-05. The boy is up to his thighs and the dog is almost buried.*

were pliable enough to put on. One learnt the lesson that it was necessary to sleep with one's boots in bed with one and, strange to believe, make them part of one's pillow.' Another soldier reported 'suffering badly from frost-bitten feet. My hands had swollen to twice their normal size and frequently my arms were completely numbed up to the elbows. For nearly a month I had not had my boots off and most of my companions were in the same miserable state as myself, owing to the intense cold... Putting the backs of my swollen hands to the barrel of my rifle, I could feel nothing, and I could only lift it up by holding it between my elbows. God help us if we had been called upon to defend ourselves. Fifty yards away [the enemy] were, no doubt, thinking the same thing. Frostbite was a terrible enemy, how terrible only those who lived in the trenches for weeks on end knew. Men were actually being frozen to death and others were fighting against sleep and cold until they could fight no longer.'

As that soldier had surmised, the German soldiers were in the same miserable state. As one, Paul Hub, remarked, 'Through all those wet, snowy, cold days, we were lying, wet through and with chattering teeth in the mud and filth of holes that we had dug ourselves in the ground.' The one benefit of the weather came when lice-ridden soldiers put their shirts out in the frost to freeze the lice to death and then warmed their shirts by sitting on them before putting them back on. Unfortunately, like all the other remedies against lice, it offered no more than temporary relief before the itching, scratching plague returned.

In 1932-33, snow fell in Scotland in October and from 23 to 26 February there was a "Great Blizzard" with gales and prolonged snowstorms, prompting the Met Office to issue its first ever warning about disruption to traffic. Whipsnade recorded 24 inches of level snow, Buxton 28 inches and Harrogate and Huddersfield 30 inches, and there were drifts of fifteen to twenty feet on the hills.

The winter of 1939-40, the first of the Second World War, was one of the longest and snowiest of the century. January 1940 saw savage cold, heavy snow and deep drifting right

*Britain's highest pub, The Tan Hill Inn, half-buried by drifts.*

*A tidal stretch of the River Ouse frozen solid in the winter of 1939-40. The harshness of the winter compounded the misery of wartime Britain.*

*Ships frozen into the river while waiting to load at Scott's Mill, Selby in 1940.*

across the country - Sheffield had a four-foot snowfall towards the end of the month. The West Highland railway line in Scotland was blocked and a train was also snowbound a few miles south of Preston for thirty-six hours. On 27 January rain fell on frozen ground over a swathe of land stretching from North Wales across the Midlands to the South-West and central-southern England. The ice remained intact for a week, lasting until 3 February, causing massive transport disruption and hundreds of injuries to people slipping and falling.

British families endured the privations of war far into peacetime, with rationing in force on some products until 1954, almost a decade after the end of the war, and in 1947 a winter occurred that piled further miseries upon them - perhaps the most harsh winter ever experienced.

*29,325 supporters enduring a blizzard at the F.A. Cup fifth-round tie between Halifax and Luton in 1933*

## PART II

# The Snowiest Winter - 1947

There have been colder winters than 1947 - though not many - but no winter since records began has ever been as snowy. It began gently enough. There had been snowfalls before Christmas and again in early January 1947, but both had melted within days and by the middle of that month, the temperature was unseasonably warm. At Leeming in North Yorkshire it remained in double figures throughout the night of 15-16 January, and the next day, it rose to a balmy 14°C right across the country, from Norfolk in the East to Herefordshire and Flintshire in the West, prompting optimists to begin looking forward to spring. It was to be a long time coming.

The weather turned colder again on 20 January, with a sharp frost that night and the next. On the 22nd, clouds the colour of pewter began to mass, building from the south-west, and it started to snow in mid-afternoon. A few flurries of sleet and wet snow fell at first, melting almost as soon as they touched the ground, but the snow continued and intensified throughout the evening and night, driven before a fierce wind. The heaviest falls were over the South and South-West of England, where the blizzard was the worst recorded since 1891; many villages in Devon were cut off. Even the Isles of Scilly had a few centimetres of snow, an almost unheard of occurrence.

As the snow began to fall, the government was in crisis over plummeting stocks of coal and food. The country was still largely dependent on the rail network for the movement of people and supplies. There were no motorways, relatively few lorries and no large-scale programme of salting and gritting the roads in bad weather. There were few convenience foods, or imported foodstuffs, very few fridges and even if food had been plentiful, most families had no money to spare for anything but necessities. A Government spokesman, Lord Henderson, chose this moment to announce 'profoundly gloomy forecasts of Britain's food position'. The fresh meat ration was cut from one shilling and twopence (six pence) to a shilling (five pence) of which the portion that had to be taken as 'corned meat' - usually Argentinean corned beef - increased from 2d to 4d. There was 'no hope' of restoring the three ounce ration of bacon and 'probably difficulty in maintaining the two ounce ration' and a reduction in the bread ration 'may be inescapable'. The one feeble consolation he could offer was on the supply of eggs, with 'a good chance that they will not be fewer than last year'.

*A lorry fitted with a snowplough struggling to clear snow from a moorland road. The driver's workmates stand ready to dig him out if he gets stuck.*

As Lord Cherwell - Winston Churchill's Personal Assistant during the War - pointed out in the angry debate that followed the announcement, 'We are faced with the stark fact that in the first year of peace we had less than in the last year of war,' and Lord Wootton complained of 'a harassing unhappiness to the housewives of the nation'. The morale of their menfolk would not have been improved by the news that a fifty per cent cut in their coal allocation was forcing brewers to announce a halving of beer production 'for at least six weeks'. Coal was in such short supply that at least one gas works closed because of a lack of its raw material and many other industries were on short-time working, and the first heavy snows over the weekend of 24 to 25 January produced power-cuts of ten, fifteen and then twenty per cent throughout the country, lasting from twenty minutes to two hours.

The onset of the winter storms could only exacerbate the problems. The temperature barely rose above freezing that day or the next, and there were further heavy falls of snow spreading northwards over the whole of England and Wales. The level snow was up to a foot deep and drifts more than fifteen feet deep blocked countless roads and railways from Cornwall to Kent and northwards as far as the Scottish border. A line of abandoned lorries and cars on Wrotham Hill in Kent stretched for two miles. Salt and grit were rarely used in those days and if roads were cleared at all, the snowploughs left behind a two inch layer of hard-packed snow. Motorists either left their cars in their garages and walked, or fitted tyre-chains. So regular and predictable was the annual post-Christmas cold snap, that many motorists kept a set of spare wheels with old, bald tyres permanently fitted with chains. When the winter snows came, they swapped wheels and kept the chains in place until the spring thaw. In every area, but especially in the industrial North and Midlands, the pristine snow soon turned darker and greyer as coal smuts and fine ash from house- and mill-chimneys drifted down. By the time the snow finally melted it was often covered with a grey-black crust.

The hard frosts covered ponds and lakes with ice, an enticing sight for children. During the weekend of 25 and 26 January, four children died as a result of 'venturing onto frozen ponds' - the first lives to be claimed by the winter weather. Two boys, aged nine and fourteen, were drowned near Maidenhead after their sledge, improvised from the chassis of an old pram, broke through the ice of a frozen pond, and six year-old Raymond Bailey and his eight year-old sister, Sheila, both drowned when the ice broke while they were sliding on a frozen pond near Newport in Shropshire.

On Sunday 26 January gales were pounding 'the East Coast and the shores of the Channel', and conditions were so bad at London Airport [Heathrow] that though BOAC services to Sydney and New York left London, 'two for Cairo' were cancelled and only one inbound flight was able to land all day. Rail and road transport was badly affected by snow and ice and in Leeds bus and tram services were also hit by the number of drivers and conductors - 260 - off work with colds and flu. By the morning of 27 January, roads in Devon were buried under ten-foot drifts, and 100 miles of Kent roads were also blocked. The Rochester to Maidstone road was covered by seven foot drifts and the Whitstable to Canterbury road also 'impassable'. The Isle of Sheppey was cut off and sledges were being used for milk deliveries in Hampstead and Southend. Two fire engines from Clacton and Colchester, called to a fire at Walton-on-the-Naze, were stopped by snowdrifts and the fire had to be left to burn itself out.

The following day, 28 January, snow fell continuously in Eastbourne for ten hours. In

*An AA Box near Merthyr Tydfil almost buried by snow drifts.*

Percy Lambert, a postman at Thixendale, wearing 'gumboots, a thick overcoat and a muffler over his head', walked his 21-mile round, skirting drifts six feet deep. It took him from 7.30 am to six in the evening, but he was out again first thing the next morning; as the postmistress said, 'His motto is: the post goes through'. Over the course of the week, he covered over 120 miles on foot through the thick snow. One of his customers, Fred Dawson, of Riggs Farm, near Thixendale, had seen 'not a soul' for a week and was further isolated as 'his wireless battery is discharged. Shortage of water is his worst trouble. There is no road through to the only supply of drinking water - a roadside tap two miles from the farm.' His stock depended on a frozen dew-pond from which water had to be 'ladled through a hole in the ice.'

Throughout Britain, soaring demand for fuel, reduced output from the collieries, problems with distribution because of the weather and shortages of locomotives and rolling stock, were combining to make an already bad situation even worse. Iron, steel and engineering works were forced to reduce production, Lancashire cotton mills and Yorkshire wool mills went on to short-time working or closed down altogether, the Abbey Paper Mills in Holywell, North Wales, shut down, and in Leeds 5,000 workers were sent home because of power-cuts. Shoe factories in Northampton were 'at a standstill' and workers at the Austin Motor Company's Longbridge factory in Birmingham were told that fuel shortages meant that until further notice, they would be restricted to working 'one day in ten or one five day week in ten weeks'. The government issued warnings that nationwide power-cuts would be imposed from seven in the morning to seven at night if the situation did not improve. To reinforce the warnings, on 30 January there were 'the heaviest cuts in gas and electricity supplies yet experienced' - twenty-five per cent. Standpipes were installed in the streets because the pipes in so many houses were completely frozen and even where water was flowing, waterworks in Cambridge and other towns and cities warned that there would be cuts in supply because of the scale of leaks from broken pipes.

On top of rationing and other food shortages, milk was now in short supply with mounting stocks of uncollected milk on the farms. Some farmers struggled across the fields, hauling churns on sledges but even then, dairy lorries often could not get through to collect them and on many farms milk hadn't been picked up for four days. In the midst of this and other food shortages, there was outrage at the owners of a Bournemouth hotel prosecuted for 'receiving 9,000 eggs during a period when it was entitled to only 1,980 - a complete black market in eggs'. The eggs had been pickled and 'stored in a room entered by a trapdoor'. Future egg-supplies throughout Britain were threatened by the death of thousands of chicks in incubators after cuts in the electricity supply. On 31 January there were even worse reductions with twenty-five per cent cuts in electricity and even more savage cuts in gas supplies - fifty per cent in North Middlesex.

Worse was to come as Britain endured the coldest February since records began in 1659. The month began with a brief flicker of a thaw - on the first day of February the temperature reached 6°C at Penzance and on the Air Ministry roof in London it rose above freezing for the first time in eight days, but by six that evening it was back below zero. The next day a twelve hour blizzard hit the North - once more 'the worst of the winter'. Six-foot drifts cut off Edale in Derbyshire and every trans-Pennine route was blocked, apart from the road over Blackstone Edge, kept open by 'the constant use of snowploughs'. The driver and conductor of a bus from Settle to Horton-in-Ribblesdale had to dig themselves out of drifts 'three times with garden spades', and the power lines to Hemsworth were brought down during Evensong at the village church. 'Candles were lighted and after tapers were distributed to the congregation, the hymn "All Things Bright and Beautiful" was sung.'

The blizzards and gales paralysed shipping traffic too. In the North-East, Seaham harbour was closed because of gales and high seas, and on the opposite coast, passengers on the Isle of Man packet steamer *Mona's Queen*, who had already been waiting at Liverpool for two days, 'last night settled down to another weary wait at the landing stages as they had been told through loudspeakers that the earliest departure would be at four o'clock this morning, and then only if the weather cleared.'

Meanwhile, HMS Vanguard, bearing the King and Queen and Princesses Elizabeth and Margaret away from frozen Britain for a six-week state visit to South Africa, was battered by the same storm in the Bay of Biscay. 'Heavy seas smashed gratings on HMS Vanguard's deck and broke adrift a piano in the ship's schoolroom', and the Royal Family spent most of the day quietly in their quarters.

The next morning, the 250 ton trawler *Elsie Cam* sank at the mouth of Grimsby harbour after it had 'battled for several hours with one of the worst blizzards the Lincolnshire Coast had known.' Police officers patrolling the sea-front 'had difficulty in staying upright against what one described as "a hurricane".' Captain Almond, the skipper of the trawler, sent a last radio message 'Just about had it. Don't think you can do anything for us. Give my love to wife and kiddies. Can't last. Goodbye. Gone.' In fact, although the trawler sank, he and his nine-man crew were all rescued from the freezing sea.

By the next day, 3 February, continuing snow had left Buxton in Derbyshire cut off by road and rail, power lines had been brought down in County Durham, 'plunging a large area around Tow Law into darkness' and traffic was 'at a standstill' in many parts of the country. The Leeds-Liverpool Canal was frozen solid at Skipton, holding up barges laden with coal. Ice on the transmitter at Moorside Edge forced BBC programmes off the air in the North, 'believed to be the first time in this country that a "hold-up" has been caused in such a way.' At Castle Hill near Huddersfield, three double-decker buses were 'marooned' in snow drifts, along with 'another bus, equipped with a snowplough, which had gone to their rescue and then got stuck itself.' Mr RC Chadwick, a postman in the Kirkburton district, 'sank up to his chin in a snowdrift and had to burrow his way out with his hands.' During the FA Cup replay at Leicester City that night, two Brentford footballers begged the referee to abandon the game because of the blizzard that was raging, but he refused, perhaps feeling that the 4-1 scoreline in Leicester's favour might have had something to do with their pleas. Late that night 'with the aid of lights from snowploughs', teams of men dug through drifts on Dartmoor from Princetown to Yelverton, about eight miles away, to enable an ambulance to take two women in need of medical treatment to hospital.

The snow was falling as heavy rain in East Anglia and there were floods in Chelmsford, Ipswich and Felixstowe, but elsewhere in Britain there was only an endless expanse of white. The next day, 4 February, the snow was even heavier - 'the worst day's blizzard ever remembered', continuing non-stop for over twenty-four hours. Forty villages in Leicestershire were cut off, the Vale of Belvoir was blocked by seven foot drifts, there were ten foot drifts at Cromer in Norfolk, and on high ground in the North the drifts reached twenty feet and more. The AA reported that 'all roads were impassable in Derbyshire' and that 'snowdrifts several feet deep' were making road travel 'virtually impossible' throughout the North

*Undertakers were forced to turn to horsepower as blocked roads prevented motor hearses from getting through.*

Midlands and the North. 600 vehicles were caught between Grantham and Stamford, and the Settle-Carlisle railway line was blocked virtually all the way through the Pennines; 'for miles the lines were indiscernible'. Four engines of the LMS railway were derailed at Low Moor, Bradford, and the roofs of Grantham and Wakefield stations collapsed under the weight of snow. At Grantham a seventy-five yard stretch of the roof caved in, completely blocking the up and down main lines only seconds after an LNER express train had passed through. A 300-foot section of the roof of the Avro factory at Yeadon in Yorkshire also collapsed under the weight of snow, fortunately a few minutes before the men arrived for work.

The misery was not confined to Britain; the whole of the Northern hemisphere seemed to be gripped by ice and snow, with a temperature of -78.7° Fahrenheit - 110 degrees of frost - at Snag airport in the Yukon, 'the coldest weather ever recorded on the North American continent'. It was reported from Berlin that fifty-three Germans had been found frozen to death on a refugee train from Poland when it arrived at Dresden.

In Britain, there were tales of lesser, but still considerable hardship from every quarter. A party of seventy-eight people, many of them elderly, returning by charabanc to Pickering in North Yorkshire after a trip to see the pantomime in Whitby were trapped overnight on the moors by snowdrifts. Some of the passengers 'fainted from the cold during the night'. A motorist who came across them 'tramped for four hours over the snow to Market Weighton, four miles away', to alert rescuers. The next morning the trapped passengers were led on foot to the nearest village, Lockton, where they were housed by villagers until the storm abated.

Another busload of sixty pantomime-goers, returning to Littondale from Bradford, struggled through the snow to the Devonshire Arms at Bolton Abbey, 'where they were accommodated in the hotel ballroom'. Three members of the party, all dairy farmers 'anxious to get home to their farms in time for milking,' then set off at two-thirty in the morning to make a seventeen mile trudge through blinding snow and deep drifts. They did not reach their farms until after lunchtime that day. Thirty-five other members of their party were still at Bolton Abbey three days later. The crew of an ambulance from the Keighley District Victoria Hospital used a sledge to carry a seventy-five year-old woman a mile and a half over the snow to their ambulance, and two doctors fought their way through nine miles of drifts in Wensleydale to rescue a nine year-old girl suffering from appendicitis. Yet, for one six year-old Leeds girl, the snow was a life-saver. The girl had been 'dancing too near the fireplace' when her 'loose frock' caught fire. She ran out of her house with her clothing ablaze and a neighbour, a sixty-nine year-old widow, rolled the little girl in the snow to put out the flames. One man already beyond any but divine help, was carried to his funeral in Holmfirth by 'a special sleigh' constructed by the undertaker, 'as it was impossible to use a hearse' because of the snow.

There was great alarm over the fate of a couple from Kensington in London. Estate agent Richard Grisham and his wife Joy had disappeared while on holiday in the Yorkshire Wolds. Despite the snow, they left their hotel at Goathland to walk to Egton Bridge and had not been seen since. They were still missing three days later, when 'grave fears' were being expressed about their safety. Police fought their way through snowdrifts of twenty feet to reach all but three outlying farms; 'The one hope we have,' a police spokesman said, 'is that Mr and Mrs Grisham reached one of these.' Meanwhile those able to obtain a newspaper were informed that 'the King and Queen and the two princesses were out on the open deck yesterday, as HMS Vanguard entered warm sunshine and calmer seas off Morocco.'

Councils were now using bulldozers and mechanical diggers to try to clear snow from blocked roads and contractors normally working on opencast coal mines were put to work with their diggers and grablines. German and Italian prisoners of war and members of the Polish Resettlement Corps were also issued with shovels and put to work - 'as soon as Polish troops at Blackshaw Moor camp between Leek and Buxton have dug themselves out, 2,000 will go to Buxton to help in snow clearance'. However, the weather continued to frustrate efforts to get Britain moving again.

*A steam train embedded in a huge drift. Aisgill, 1947.*

Virtually every commodity, particularly coal and foodstuffs, was in desperately short supply. The movement of coal by land and sea was badly hit by the blizzards and gales, and gas and electricity supplies were constantly threatened by the shortages of fuel. Thirty-seven vessels, fully loaded with coal, were trapped by the gales in North-East ports and a further thirty, empty, were immobilised in London. It would have required 150-200 coal trains to move the fuel by land, but there were no spare trains available and in any case, most lines were blocked by snow.

Virtually every Pennine town and village was running out of bread and food, and with Buxton still cut off and desperately short of supplies, 'lorries laden with fresh meat' and other supplies were waiting at New Mills to 'rush in food' as soon as roads could be opened. Even in the lowlands, the situation was often desperate. A woman in Warrington, who had no fuel and was also running out of food, remembers pushing a pram four miles through the snow to collect a sack of coal and a loaf of bread. Dairies in many places were now delivering milk only to those with "priority permits" - nursing mothers, hospitals and schools - and a spokesman for the Craven Dairies in North Yorkshire doubted if there was even enough for them. There was a glut of uncollected milk on farms and 'hand-turned churns were being used for the first time in years' to turn some of the milk into butter before it went sour.

By 6 February, 60,000 workers were idle in Birmingham alone and the government was forced to impose 'drastic restrictions on railway and sea-borne traffic to relieve the critical coal supply position caused by blizzards on land and gales at sea, as the tally of industrial concerns forced by coal shortages to close down continues to grow.' The Ford Motor Company at Dagenham was the latest to shut down, 8,000 workers at Cadburys in Bourneville were sent home for ten days and 2,000 workers at the Goodyear tyre factory were also laid off 'for an indefinite period'. Two Co-operative depots in Sheffield, supplying sixty per cent of the city's population with milk, had to send their workers home because they had no milk at all.

*On some trans-Pennine roads, vehicles were stranded for weeks.*

A ban was imposed on the export of coal, railway companies were ordered to 'curtail general traffic to expedite the movement of coal at all costs,' and power stations were given 'prior call' on coal stocks, though 'domestic supplies would not be touched, except as a last resort.' Efforts to improve coal production and distribution were hampered by a series of problems. Many families reliant on open fires as their only source of heating had already used up their coal and could obtain no fresh supplies as coal merchants were snowed in their yards. Eight pits in Nottinghamshire and Derbyshire and two in Northumberland had to close because of 'a lack of coal wagons', colliery sidings in the North-East were blocked by snow and the coal docks at Barry and Cardiff were paralysed by the 'freezing up of loading devices'. One industrialist, GB Warman of Shenstone Air Developments in Lichfield, shut his factory and issued an appeal to all his workers to 'volunteer to go down the coal mines in this country's hour of need.' The response was 'so gratifying and, he admits, amazing, that he has passed on the idea to Mr Lord, General Manager of the Austin Motor Company at Longbridge, where 15,000 workers are now idle.' 'The relief of coal-starved London' began on 7 February with the arrival of seven colliers off Tilbury, carrying 18,000 tons of coal.' Meanwhile 'a concentrated attack' by bulldozers backed by an army of hundreds of men with shovels, opened several main roads through the Pennines.

The search for the Grishams, the Kensington couple missing somewhere on the Yorkshire Wolds, was called off the same day after a telegram handed in at Jermyn Street Post Office in London was received in North Yorkshire. Without bothering to inform anyone, the Grishams, it seemed, had simply gone home, leaving their luggage to be collected at some later point. A police spokesman condemned 'the thoughtless behaviour of these people that has caused tremendous expense to the county and risk and trouble to the police, special constables and other volunteers' who had taken part in the search for the supposedly missing couple.

It would be pleasing to think that the Grishams were at least inconvenienced by the snow that returned to London on 8 February, with a storm lasting right through Saturday night. Five to ten inches fell on the capital and, 'owing to the lack of men, for the first time within memory' snowploughs were used in the City of London, and at Hornsey, Muswell Hill and Highgate. People were out 'skiing and tobogganing' at Boxhill. The blizzards swept on over the country, isolating many Shropshire villages and cutting off 'an area of 300 square miles' north of Market Rasen in Lincolnshire. In the uplands, ramblers and rock-climbers volunteered to help save buried sheep flocks, and the members of an RAF mountain rescue team, each carrying thirty pounds of food and 'guided by compass and radio', reached the village of Huggate on the Yorkshire Wolds, cut off for eight days. It took them three hours to cover just two miles.

On Monday 10 February, day-long power-cuts were imposed, leaving much of London and other towns and cities working by candlelight, hurricane lanterns and 'the grey winter light that filtered through the mist'. Many shops, cafes and restaurants closed, there were cuts in tram and train services and escalators in department stores and the London Underground were switched off. Most London theatres cancelled their matinees and the BBC announced that the Third Programme and all television broadcasts would be suspended 'for the duration of the emergency'. Several periodicals were ordered to cease publication until the power situation improved, and whether through altruism or in the hope of poaching their star columnists, some of the daily newspapers offered the periodicals a modest space within their pages. In the City, Whitehall and the West End, the power was left on, but a request for voluntary cuts 'appeared to be scrupulously observed', except at the Bank of England, which had its own generator, and the War Office, where a disapproving reporter noted that 'lights blazed from nearly every window'.

There was a brief thaw in the South that day but by Tuesday 11 February, temperatures had again sunk below zero and the snow was falling once more. This time the East Coast bore much of the brunt. The Great North Road was again blocked north of Alnwick, Cromer and Sheringham in Norfolk were cut off by six-foot drifts and all train and bus services in North Norfolk were brought to a standstill. No fish trains were able to leave Grimsby and 200 prisoners of war had to dig out a trainful of passengers for Peterborough stranded for twelve hours at Ludborough. Other trains were 'embedded in snowdrifts' near the Yorkshire coast. Fifteen men hauling a sledge battled through the snow to an isolated farmhouse at Settrington, East Yorkshire, from where they retrieved the body of a farmer, who had been dead for more than a week.

Second World War Halifax bombers were brought back into use, flying from Fairford in Gloucestershire to make food drops to Leek and Longnor in Staffordshire, where there had been no food deliveries for ten days. Villagers at Longnor marked out a 'black cross of soot' to help the pilot locate them. On 13 February, another mercy flight from Fairford to Butterton in Staffordshire ended in tragedy when the Halifax bomber making the food drop crashed on Grindon Moor, eight miles from Leek, killing all eight crew.

There was a brief respite from the snow around the middle of the month, but temperatures plunged even lower; 17 February was the coldest day in Falmouth since the nineteenth century. Councils and train companies seized the chance to clear blocked roads and railway lines. In the Northern Command area, 2,610 troops, together with 2,635 German POWs and 2,120 Poles were at work snow-clearing. In Bristol a thousand workmen laboured to clear the streets; as they finished, it began to snow again. Once more the snow spread right across the country, again closing roads that had only been cleared 'after days and nights of work'. In the Rossendale valley, where the snow was 'up to the level of the windows', some roads were blocked for five weeks. In Scotland, water supplies were threatened in Stirlingshire because the feeder streams to the reservoirs were all icebound.

At sea, ice-floes from the Rivers Scheldt and Rhine were sweeping along the Belgian coast, freezing together between Knocke and Blankenberghe into an ice-shelf ten inches thick and extending well out to sea. People walked on it up to 300 yards from the shore. Ice-fields had formed in the North Sea and drifted to within forty miles of the Norfolk Coast. Two Fishery Protection vessels sailed from Lowestoft on 18 February to investigate. They returned the following day warning that the ice was still drifting westwards and formed 'a menace to shipping' both in itself and because the ice was carrying away buoys marking wreck-sites and shipping channels between areas of uncleared mines from the war. A huge floe was also reported east of Great Yarmouth and over the next few days the ice drifted steadily closer to the coast. A Ministry of Agriculture and Fisheries official revealed that the sea temperature was two degrees below freezing point, and said that 'there is no record of anything like it being known before.' There were fears of 'much destruction' of stocks of herring and plaice because of the cold. Two trawlers were trapped in the ice before managing to escape and the Southern Railway announced the suspension of the Ostend ferry service 'until further notice' because of the danger from ice-floes.

*31 January 1947: The milk must get through. A milkman in Cheam, Surrey, uses a sledge to make his deliveries.*

## Snowed up in the Isle of Wight.

*At teatime this afternoon, before we had drawn the curtains, I looked up from the Times crossword puzzle to see that while there was still a patch of pale cerulean high in the south-east corner of the window, everywhere else it was snowing. Instantly, just for a moment, I was restored to the magic world of childhood. Magic and childhood apart, I tend to associate snow with my youth, when I lived among the hills of the West Riding [of Yorkshire], and winter came much earlier - and much harder - than it does now, and three years out of four there would be heavy falls of snow around Christmas. (If we are all deceiving ourselves about this, then why is there so much snow on Christmas cards?) Even so, the only time I found myself completely snowed up was as late as February 1947. I was living in a fairly remote old house called Billingham Manor (said to be haunted) in the Isle of Wight. There was no nonsense about this snowed-upness. The one road to and from the house was blocked, utterly impassable; nothing could be delivered; the telephone lines were down; and for about a week we were cut off, sealed off, lost to the world. I passed the time writing a play over which I had been brooding for a month or two, and I had finished it when the road was clear again. It was called The Linden Tree and, with Sybil Thorndike and Lewis Casson leading the cast at the Duchess Theatre, it was an immediate "smash hit". But some of the credit must go to that astonishing snow storm in February.'*

**JB Priestley.**

*Beautiful but deadly – hoarfrost cloaks every twig and grass-stem.*

*A snowstorm sweeping Oxford Street in London. Normally one of the busiest streets in the world, it is eerily deserted, as shoppers once more stay at home, to the consternation of store owners.*

UPPER PART
TO-LET

*A tug and a barge frozen into the ice of a river. Tugs were used to try and keep waterways open, but they too sometimes became caught in the ice.*

19 February was the eighteenth consecutive day without sunshine at Kew, the longest sunless period ever recorded - a record that was to be extended to twenty-one days. The sun 'penetrated the clouds' briefly around noon, but the feeble light through the mist and murk was 'not strong enough to record on their instruments'. Even when it did eventually break through, it shone for a total of just seventeen hours in the whole month.

The next day, 20 February, the snow returned once more, spreading from the South-West to cover the whole country and continuing for most of the next two days. By the time the storm ended, it had left 'whole trains buried under drifts'. Miles of telephone and power cables had been brought down and many telegraph poles collapsed under the weight of snow, or of ice which was often two inches thick on the wires. The Settle-Carlisle railway line was again blocked by twelve-foot drifts and men laboured in vain to clear them.

Once more temperatures were in freefall, and a minimum of -21°C was recorded at Woburn in Bedfordshire early on 25 February. At Dungeness, Eastbourne and many other points around the coast, the sea was frozen up to one hundred feet from the shore, and at Whitstable in Kent, hundreds of people crowded the beach to watch 'masses of ice-floes extending for miles along the shore and over a mile out to sea'. In the harbour, barges and fishing vessels were icebound and as the tide ebbed, huge blocks of ice were left on the mud-flats. Pack-ice had formed in the estuaries of the Solent, Mersey and Humber. 'Practically the whole of the western seaboard of Wales is frostbound, and in the upper reaches of several North Wales rivers, ice-packs, some up to two feet deep, are floating downstream'. The Avon froze at Bristol for the first time in sixty years. People skated on the frozen lake in St James's Park in London, within sight of Buckingham Palace. Skaters were also out on the river at Evesham and the River Dee was icebound over a length of two miles. The Thames again froze at Windsor, and ice covering the water above Romney Lock had to be broken up. The Grand Union Canal was covered by a thick ice-sheet from Willesden to Paddington, leaving strings of barges icebound, and the ice on the Medway at Rochester was so thick that icebreaking tugs from the Admiralty had to be used to open a channel.

*Two trains snowbound near Ribblehead on the Settle-Carlisle railway line*

For man and beast food was proving ever harder to find. Already scarce under rationing, many food items began to run out altogether. Supplies of milk, butter, meat, fish, and even bread were perilously low. To alleviate some of the shortages, the government had imported ten million tins of South African snoek, a fish hitherto unknown in Britain. Despite valiant attempts by Whitehall propagandists to convince the populace of its virtues, including the distribution of recipes for dishes like "Snoek Picquante", most were resolutely unimpressed.

Fresh food in shops, barns and warehouses was frost-blighted and winter crops like cabbages, sprouts, parsnips and turnips were buried feet deep beneath the snow and often frozen into the ground. Some farmers, like JH Moorhouse in Essex, even used pneumatic drills to dig parsnips out of their frozen fields. Farmers remained unable to plough their fields and plant their crops for three months, guaranteeing that the 1947 harvest would be one of the worst of the century

Feed for livestock was also running out in many areas, and the losses of hill sheep were at catastrophic levels, 'higher than ever experienced'. Sheep buried by snow can last for days and even weeks, even eating their own fleeces to survive, but many died of starvation and cold, or were suffocated by wet snow when a thaw at last came, and those that were found alive were so weakened by their ordeal that huge numbers lost their lambs and many died themselves. 'Sheep are wandering over walls and have eaten the bark off trees. They stand about the porch of the village school waiting for scraps given them by the children. Two stray sheep which came in to feed with one flock were later found dead, one frozen to the barn door, the other with its head frozen into a watering trough where it had been to drink.' One farmer's cattle were so weak from lack of fodder that he 'had to lift them up each morning to feed them. Some households were near starvation point themselves.'

Over a thousand sheep were 'dead or dying on the moors around Ribblehead, North Yorkshire, and three farmers each lost over half their flocks. A farmer from Mallerstang in Westmorland's upper Eden Valley, only moved onto his farm in the autumn of 1946, purchasing the stock of 500 ewes, all of which would have been carrying at least one lamb by the time the snow came. By the end of lambing time the following Spring, there were only 170 ewes and twelve lambs still alive. Some sheep in Lincolnshire had to endure a further hardship, for 'flocks of hungry crows are attacking sheep and some have been killed.' In the Lake District too, shepherds were 'now armed with guns to combat flights of carrion crows which are the latest menace to mountain sheep. Deprived by snow of their natural food, dozens at a time are swooping down on helpless sheep trapped in drifts, first tearing off their ears and plucking their eyes, then devouring the whole carcass'. Farmers in Merionethshire also reported large numbers of foxes 'driven by hunger from their mountain lairs to the lowlands, making daylight raids on sheep'. Around two million sheep were to die during the winter, 500,000 acres of wheat was lost and the frosts also destroyed much of the late potato crop, promising yet more hunger and misery for British families.

On the night of Tuesday 25 February yet another massive snowstorm swept the country. Northern Ireland, Eire, Scotland, Northern England, North Wales and the North Midlands were hardest hit. Scotland and England were completely cut off from each other as snowdrifts blocked every single cross-border road. The sleeping car express from London took ten hours to travel from Newcastle to Edinburgh. At one point the engine had to be uncoupled and 'by pushing and reversing, ploughed through ten foot drifts'. Newcastle and Sunderland were almost completely cut off. Two Greek steamers ran aground in the blizzard

*Two men are dwarfed by a huge drift in front of their farmhouse.*

*"No School Today". A small child trying not to look too disappointed at the news.*

at Cullercoats and Whitley Bay in Northumberland, and the Belfast mailboat wrecked on the Isle of Arran. An avalanche sweeping 3,000 feet down a Merionethshire mountainside 'buried a farmer and his son in bed as it engulfed their farmhouse'. Their Dutch barn was carried 300 yards down the hill by the snow. By the time the storm blew itself out, there were drifts of twenty feet and more in Upper Weardale and North-East Durham, and all the surrounding dales.

On the last day of the month there was another 'very hard snowfall' in Kent and in Southend it snowed 'practically all day'. Villages on the Yorkshire Wolds were beginning their sixth snowbound week. Maximum temperatures had remained close to or below freezing throughout the month; in Oxford temperatures were below zero from 6pm on 10 February to 6am on the 26th. At Kew, chosen as the site of the Royal Botanic Gardens because of its gentle climate, on only two nights in the whole month did the temperature remain above freezing point. Deepening the winter gloom still further, Kew, Nottingham and Edgbaston all recorded no sun whatsoever on twenty-two of the month's twenty-eight days, and in some parts of Britain, snow fell on all but two days that month. Even in normally mild Dorset, there were only two nights without frost between 16 January and 11 March. Yet in a bizarre reversal, some places that would normally have expected the worst of winter weather had no rain or snow at all. A completely dry month in western Scotland is unusual at any time; in February it was unheard of, yet while the rest of the country suffered in the snow, in the West of Scotland the whole of February 1947 was dry and sunny.

*Just a group of three men, pausing from their snow-clearing duties... except that the snowdrift they're standing on reaches almost to the top of the telegraph pole behind them.*

For the rest of the country, commerce, industry and normal daily life were paralysed. In many areas people could only leave their homes and travel to their towns and villages by walking on the top of the snowdrifts, which by now were level with or higher than the wall-tops and hedges, and the tops of signposts and telegraph poles were often the only visible objects in a wilderness of white. Industrial production had fallen by a quarter, unemployment had reached two million and another two million were temporarily laid off because of power-cuts.

Despite the mountainous snowdrifts and arctic temperatures, valiant - and usually unsuccessful - efforts were still being made to break winter's grip. At Silloth in Cumbria, the stables at one farm were completely covered with snow, and the farmer could not get his horses out for two weeks. He had to leave his farmhouse by his bedroom window and then dig down through the drift in order to feed his horses. As he and his sons began trying to clear the snow from his gateway, an RAF snowplough came along the lane, but the snow was so thick that, after taking a run at it, the plough bounced back, leaving the driver dazed but uninjured. 'It is thought that, after limping back to Silloth Aerodrome, the plough never moved again.'

Repeated attempts were made to clear the railway line over Stainmore between Bowes in County Durham and Kirkby Stephen in Westmorland, a vital link for farmers and their isolated communities, and between the Durham collieries and the mills of industrial Lancashire, blocked for over a month. The biggest obstacle was a snow-filled cutting at Bleath Gill, near Barras Station, just below the summit of the line. Attempts to clear the drifts using a train-driven snow plough failed - the plough became embedded in the snow and had to be dug out by hand. Gangs of hapless Italian prisoners of war, already held in miserable conditions in a disused Army camp on a windswept moor a few miles away, were compelled to dig out the cutting with picks and shovels, but every night there were fresh snowfalls and the wind also blew the excavated loose snow back into the cutting. After a fortnight's backbreaking work, they were farther away from the head of the cutting than they had been when they started.

Rail and Ministry of Transport officials then had the idea of using one of Britain's newest and most celebrated inventions - Sir Frank Whittle's jet engine - to clear the snow. Two powerful jets 'capable of developing 2,000 pounds pressure [each] and releasing hot gases at 1,000 miles an hour' were mounted on the rear truck of a train and backed up to the blocked cutting. The turbines began to revolve and the whine of the engine grew to a scream. The 'effects of the attack by these jet engines is spectacular. The blast sends up a huge cloud of fine snow particles, accompanied by great blocks of frozen snow which shoot up to a height of about thirty feet. There was a fear that damage might be done to telegraph poles and wires along the track, but so far this has not happened. The assault, which at the moment is still experimental, opened on Saturday and would have been maintained throughout the night, but in the exceptional cold of Stainmore Summit, 1,370 feet above sea level, the electrical equipment of the jets froze and night work had to be abandoned.'

Initial reports of progress were optimistic. 'Already the progress made by these jet engines is equal to the thousands of man-hours worked by hundreds of troops and LNER relay gangs during the past month, in what must have been the worst blizzard conditions in the country.' However, 'ahead lies the worst part of the job. There, deep cuttings through rock have been filled, often to a depth of thirty feet. A train is buried; it had to be abandoned a month ago.'

*Sir Frank Whittle's new invention - the jet engine - being used to clear snow from railway lines. Although effective against soft, new-fallen snow, jet engines made virtually no impression on the hard-packed, frozen drifts blocking railway cuttings in the Pennines*

*A train using its snowplough to blast its way through thick snowdrifts. It looks like an explosion – and dynamite was often used as the only means of clearing hard-packed snow in the worst affected areas.*

MXT 234

Starving sheep were reported to be 'squatting' on the railway line near Goathland in Yorkshire and train drivers and firemen had to climb down from the footplate and 'push the sheep aside'. The postman had not been able to reach the 'seven cottages and eighteen inhabitants' of the remote Yorkshire Dales hamlet of Oughtershaw in Upper Wharfedale since 1 February and the fresh snowfalls meant that their isolation would be further prolonged. The only contact with the outside world had been achieved by villagers who had battled through the snow to Hubberholme and Buckden - several miles away, where food had been left for them by 'tradesmen who could penetrate no further. The goods were carried back to Oughtershaw 'on sledges and by hand' through sixteen-foot drifts. Despite those supplies, one of the villagers, Mrs EJ Speed, warned that 'the food position for both people and animals is serious. We were compelled to fell a tree in order to keep warm. We have no telephone, are seven miles from the nearest doctor and twelve miles from the nearest nurse.'.

Temperatures plunged still further and there was one of the coldest March nights ever recorded with lows of -21.1°C in Braemar, Peebles and Houghall. The 'terrific chaos' continued into the next day when further heavy snowstorms caused fifteen foot drifts at Worcester, stranded more than 500 vehicles between Cheltenham and Burford and a further 300 between Cheltenham and Cirencester, and cut off a hundred villages in Leicestershire and Northants. By five that afternoon, not a single London to Birmingham route was passable and once more, said the AA, it was 'almost impossible to get from South to North'.

There were seven hour delays on the London to Brighton line and a train limped into Euston Station, having taken twenty hours to travel there from Wolverhampton - it would almost have been quicker to walk. Buses were stranded halfway up the hills at Wimbledon, Putney and Crystal Palace, Rutland and Oakham were cut off and there was twenty-four hours continuous snow in Bedfordshire. At Brighton, eighty-three snowploughs and 900 men were 'employed all night clearing snow', and in Liverpool, where snow fell for twenty hours non-stop, 1,400 men worked all day trying to clear the city centre.

Passengers on a Manchester-bound train 'embedded' in a snowdrift at Tonfanau, Merionethshire, were taken to a nearby army camp for the night. Sailings from Holyhead were interrupted and 'the *Princess Maud* was fifteen hours late sailing'. At Whitby on the Yorkshire coast, spectacular icicles formed on the cliffs now 'reached down to the sea at high

*An AA man uses his radio telephone to call for reinforcements.*

tide' and in the Yorkshire Dales, the Askrigg to Carperby road was reopened for the eleventh time that winter and then immediately blocked again. Elsewhere, the movement of coal was once more at a standstill and collieries in the Rhondda and West Wales shut down as absenteeism reached thirty-five per cent at some pits. It was even worse at Cannock Chase in the West Midlands, where fifty per cent of the miners were absent.

By the next day a huge swathe of Britain running north and east from Devon to the Scottish borders was completely 'snowed-up'. Attempts to use modern technology in snow-clearing again had mixed results. The 'use of heavy tanks swaying from side to side to break up the ice-top on snow drifts met with only limited success' and the 'use of a jet engine mounted on a tank in Derbyshire was of little avail against tightly-packed, frozen snow', though a Valentine tank fitted with two jet engines was successfully used to clear ten-foot drifts in Leicestershire.

The coal and power supply situation remained critical and in a bid to safeguard supplies, a Bill to extend British Summer Time until 2 November and reintroduce "Double Summer Time" from 13 April to 10 August had been speedily passed through the House of Commons on 5 March. Home Secretary Chuter Ede told the House that, although the move would produce 'almost negligible' estimated savings of 120,000 tons of coal in generating stations and 10,000 tons a year in gas works and domestic supplies, plus a further 20,000 tons from the reintroduction of Double Summer Time, it was necessary to spread the load on power stations and 'flatten' peak electricity demand, until the generation of power could be increased. The Bill would also make 'double dayshift working' possible.

There was further heavy snow in Yorkshire on 7 March when thousands of roads throughout the country remained completely blocked, and on 10 and 11 March, although much of the Midlands and the North had its first frost-free night for seven weeks, southern Scotland had its heaviest snowfall of the entire winter. The blizzards spread to the Scottish Highlands, where drifts of more than

*A blizzard on a Glasgow street brings trams and buses to a grinding halt*

*Mountains of snow and ice ploughed from the roads were piled up at the kerbsides to await the spring thaw.*

Longhurst 'where two goods trains were embedded in deep drifts'. 600 passengers making for Scotland spent the night at Newcastle, 'curled up in railway carriages', while gangs of railway workers struggled even to clear a single track between Newcastle and Berwick. By the afternoon, enough of a way had been opened for two trains, preceded by a snowplough to have 'pushed their way through to Berwick, their passage being guided by hand-operated signals'. The line was then closed again and 'given over to the engineering section', which then had the task of clearing the up-line on which 'snow blocks remained so great that two goods trains, an empty coach-train and a snowplough were still buried in drifts'. The 5.10 from Glasgow to Stranraer was embedded in a thirty-foot drift thirteen miles from Stranraer, where it remained for two days, and the Newcastle-Carlisle line was also blocked by two derailments in the snow, one a coal-train, the other a goods train, causing 'the worst main-line hold-up on record in the North.'

Records were also tumbling on the roads, where the 'biggest road transport hold-up ever known' was occurring on Shap Fell in Westmorland. 'The worst section of the England to Scotland main road has a seven mile succession of the biggest drifts of the century and is likely to remain blocked for several days'. £1 million of machinery and supplies was held up in the jams and the towns of Kendal and Shap 'at each end of the impasse' were 'choked with hundreds of heavy lorries.' Police in neighbouring counties were stopping all vehicles from using the Shap route. Westmorland police advised drivers of stranded vehicles to 'go home by train, but while hundreds have abandoned their vehicles, many have received instructions from their employers that they must stay with their lorries to guard the loads.' As they sat shivering in their snowbound cabs, they would not have been amused to note that British Summer Time began that night.

Upland farmers, who had already lost huge numbers of sheep faced even further losses, with even those animals that had been rescued from drifts threatened by starvation. 'Convoys of lorries' stood ready to take fresh stocks of hay to farmers in Cumberland, Westmorland, Yorkshire, Durham and Northumberland, as soon as roads could be opened to them and one Bradford farmer, ID Grant of Wadlands Farm, Farsley, made a 300 mile round-trip to Norwich to buy four tons of hay for his stock. The journey took twenty-four hours, 'a hard trip,' said Mr Grant, 'but it was worth it; we could not see the sheep starve.'

It was almost winter's last flourish; the last widespread heavy snowfall in the North was on 15 March, though there were sporadic bursts until 17 March, by which time, snow had fallen somewhere in the UK on every single day since 22 January. 1947 had proved to be the worst winter of the century and perhaps the worst ever. No winter since records began has combined temperatures down to -21°C with such prodigious and relentless snowfalls. In three months a total of 210 cm of snow fell - well over six feet of level snow - and gales had raised drifts over forty feet deep in places.

As the last snow was still falling in the North, a dramatic change was already taking place in the South-West. A thick fog at dawn on the morning of 10 March indicated a sudden change in the weather. After three months of northerlies and easterlies, the wind had swung into the west and then the south-west, and mild air with a temperature of 7-10°C began edging over the frozen snowfields, its arrival signalled by dense fogs that reduced visibility to less than ten feet in places. The fogs were followed by heavy rain - on 10 March over an inch of rain fell right across the South from Cornwall to Sussex and temperatures rose rapidly with 10°C recorded at Penzance and Falmouth.

W. F. THOMPSON

The effects of such warm and wet weather on the massive accumulations of snow were entirely predictable. Rainfall and meltwater could not penetrate the still-frozen subsoil and ran off in torrents. Streams and rivers rose rapidly and, unable to cope with the massive volumes of meltwater, soon burst their banks. The weather was now inflicting a different form of torture and after thirty-six hours' torrential rain in Brixham, South Devon, boats were used to rescue people marooned on the upper floors of their houses by floods. By the evening of 11 March, vast areas of southern England were under water. As the warm air continued to spread northwards and eastwards, the Rivers Severn and Wye, swollen by meltwater from the Welsh mountains, flooded Herefordshire and Gloucestershire. The flooding spread steadily eastwards, tracking the thaw, and by 13 March, the rivers of East Anglia were also close to bursting their banks.

Landslides added to the problems. A large section of chalk cliff fell onto a bowling green at Ramsgate and a landslide at Buriton delayed trains from Waterloo to Portsmouth. Owners of farms and cottages in parts of Wirksworth, Derbyshire, were warned to evacuate their houses because the 'entire hillside above the town was moving over a distance of a quarter of a mile, after several landslips due to the effects of snow, severe frost and the subsequent thaw on the clay bed.' Despite this, 'most of the people refused to obey the warning and remained in their houses.' On 16 March a deepening Atlantic depression swept in, bringing storms, thunder and lightning and more torrential rain. There were sixty mile-an-hour gales over southern England with gusts peaking at over 100 miles an hour, damaging buildings and whipping the surface of the flood waters into waves like storm seas. By a cruel irony, in the midst of these oceans of water, over a million people in East London were without drinking water because of flooding at the Lea Bridge pumping station. Hundreds of houses were flooded as water spread over Hackney Marsh, Leyton and Walthamstow.

The gales killed at least fifteen people, including a man and wife killed in Abbey Road, St John's Wood in London, when their house 'was blown down'. Mr Edward Rotherham of the Post Office, Withybrook in Warwickshire, was killed when a tree fell on the car in which he was travelling with his wife and two small children, who were seriously injured. The top floor of a three-storey shop was blown off in Bristol, sending 'several tons of debris crashing into the street'. 'Suits were swept along a Birmingham main street after the shop front of an outfitter was blown in', there was 'extensive damage' in Bath and many windows smashed in Nottingham. A dredger broke adrift from a tug at Falmouth and went aground on St Just Point.

Large parts of Dublin were under water after 'a day of heavy rain which ended in a two-hour snowstorm'. 'Considerable damage' was done to 'fruit plantations' in the Vale of

*A horse-drawn cart carrying bales of rags is the only traffic on a snowbound Glasgow street.*

*March 15 1947. Five hundred houses and many other buildings were cut off when flood waters burst through the banks of the Great Ouse river at Bedford and swirled through the centre of town. Wagons carrying badly-needed coal were stranded in a siding.*

Evesham, the Severn valley was flooded for forty miles from Caersws to Shrewsbury with the floodwaters two miles wide in places. The Thames Valley was experiencing its worst floods in fifty years and vast areas were marked by 'unpathed waters, undreamed shores'. The waters in Oxford were 'gradually creeping into the city centre', Windsor station was closed because the tracks were submerged and the roads into the town were all flooded, with 'unbroken waters from Windsor to Weybridge'. Parts of Hampton Court Palace were under water, Eton High Street was flooded and Eton College closed. A mile and a half of the main road into Maidenhead was under water up to five feet deep and two Sherman tanks were used to tow 'from somewhere along the road, another tank which in turn had gone to someone's rescue the previous night and become immobile in soft mud.'

The River Nene burst its banks at Northampton, and at St Neots in Huntingdonshire the gas works was flooded and all supplies to the town cut off. At its worst, the flooding was ten inches above the previous highest level, recorded in 1894. In East Anglia, the dykes collapsed, flooding most of Fenland. Troops were 'drafted into the Fen country for anti-flood work', but they could do little to stop the relentless flow of water through the breaches, and river levels continued to rise ever higher.

*Householders in Trowbridge, Wiltshire wave from the top windows of their houses after the River Biss, swollen by melted snow, burst its banks and flooded the town.*

*Floodwaters cover the land near Chester. Only trees and the railway track stand clear of the flood but the rising waters bursting through the narrow arch of the tiny bridge, threaten to cut the line at any moment.*

On 18 March, the Severn was fifteen feet six inches above its summer level at Worcester, only nine inches below the record 1886 level, and still rising. Shrewsbury was 'completely encircled, with only one high-level bridge above water', and milk, bread, mail and newspapers were being 'delivered by punt'. The Thames was now flooded from Windsor to Walton, six feet deep in places, and Staines, Wrathbury, Runymede and Datchet were all cut off by road and rail. Mr R Gray of Mixnan's Farm, Chertsey, drowned when his boat overturned.

Further north, the Trent burst its banks at Nottingham, flooding many houses to first floor level. When the floodwater reached the tidal reaches, it was held back by a Spring Tide, flooding the whole of the lower Trent valley. The River Derwent overflowed its banks 'all along its course from Malton to the point where it joins the Ouse, a distance of fifty miles', and thousands of acres were buried under water five and six feet deep. The Rivers Aire and Calder were joined by flood waters at Castleford in Yorkshire, and at Mexborough, a lorry driver, Thomas E Wakefield, 'stripped and swam one hundred yards in the flood waters of the River Don to rescue a dog from a kennel in which it had been swept downriver'.

While the lowlands were enduring floods, much of the high ground in the North was still snowbound. The road from Ingleton to Ribblehead in North Yorkshire was 'still blocked though it has been reopened fourteen times', and a cutting between Ribblehead and Dent on the Settle-Carlisle railway was 'completely filled by a block of ice thirty feet deep'. Only one road into Scotland was open and between Keswick and Grasmere in the Lake District 'police are not permitting any vehicle to proceed alone'. However there was relief for a number of railway travellers snowbound for five days on their way from Glasgow to Stranraer, who at last reached their destination.

By 19 March, apart from main roads in Kent, there was no county in England and Wales where roads were not affected by floods, snow or fallen trees. Gloucester was now flooded as the swollen waters of the Severn worked their way downstream. The depth of twenty-four feet seven inches was already 'four inches above the Great Flood of May 1886', and the record of twenty-four feet eleven inches, recorded in 1852, 'may be equalled today or tomorrow'. By the next day it had duly risen to a new record level of twenty-five feet, and it peaked a further three inches higher. Tewkesbury and Leominster were now isolated and all roads from South Wales to Birmingham were impassable. The River Wye at Hereford was at its highest for 155 years -'eighteen feet six inches in the St Martin's District' while 'a gauge below Hereford, erected to deal with any emergency up to twenty-one feet, is completely submerged. Hundreds of people are without water or gas supplies and hot meals are being taken to them from a British Restaurant'.

The Thames was 'well above' its record level. 'Inscribed on locks and walls of riverside inns and houses is the date 1894; when the waters have finally subsided, new record levels will have to be marked for this year. The AA reported that 'of six main arteries radiating from London, only the Dover and Portsmouth roads were passable throughout their length. However, the floods did bring occasional advantages to the police. Three men escaping in a car after 'a smash and grab raid at Camberley' were hotly pursued by a police car which 'took up the chase at Staines Bridge, until at Hythe Lane, the fugitive car was stopped by flood water. Three men jumped out and dashed into the water', but two of them were at once apprehended.

Flood levels in the South and West began to fall after 20 March, but flood debris was still causing 'serious interference' to trains on London Transport, and rivers were still rising in Yorkshire and Eastern England. The Wharfe, Derwent, Aire and Ouse had now all burst their banks, inundating a huge area of southern Yorkshire. The Derwent was flooded for 'sixty miles from Ryedale to Wressle' and 'rose another foot at Stamford Bridge'. Mr Evelyn Walkden, MP for Doncaster, rushed to the House of Commons with a telegram he had received from the chairman of Bentley District Council: "Bentley flooded. 1000 people isolated without food for 24 hours. Small boats useless owing to powerful currents. Power-driven boats essential for food distribution and evacuation of people. Help must be here Thursday morning. Position desperate." At midnight that night, the War Office announced that two Army "Ducks" - amphibious vehicles - were on their way to Bentley from Aldershot.

Over one hundred square miles of Fenland were now under water and Spring Tides were having 'a detrimental effect'. On the roads, lorries laden with salvaged furniture travelling away from the flooded areas, passed 'WVS and Salvation army canteens coming into the area' to provide food and hot drinks for those stranded or at work fighting the floods. The bells of Crowland Abbey were rung as a flood warning when a forty yard-wide breach occurred in Deeping High Bank. The Lower Trent flooded thousands of acres around Gainsborough. People worked by 'car lamps and searchlights' in an unavailing attempt to stem the waters rushing through the ever-widening gaps in the banks of the Ouse. At Ely an attempt was made to close a 150 foot gap in the banks with a 200 foot Bailey Bridge, but 'the idea has now been abandoned'.

A thousand men were at work sandbagging parts of the Ely to King's Lynn road to prevent '20,000 acres of rich agricultural fenland being flooded'. The effort was largely unsuccessful; on 23 March 'another 12,000 acres of Fenland were flooded when a culvert between Ely and King's Lynn burst', and the next day 'flood waters west of Ely were pounding in waves' against the banks of the New Bedford River and undermining them'. On the highest of the Spring Tides at Boston in Lincolnshire, 'a 25-foot wall of seawater in the tidal estuary held up the fresh water. For the next two hours, hundreds of workers fought ceaselessly to buttress the danger spots'. Meanwhile, a ship was 'brought round from King's Lynn to block a breach in the bank of the River Wissey.' 'The spreading waters have already reached ground floor windows of hundreds of houses and many bridges have collapsed. Most of the cattle have been rescued but it is feared that potatoes and cattle feed will be destroyed.'

In Yorkshire the floods were the worst in 250 years and in York the worst ever recorded. On 24 March the river was at seventeen feet above the normal level - six inches higher than the floods of 1894 - and still rising, and the city was 'in danger of losing its supply of water and electricity'. Townspeople were warned to boil all water. Downstream from York, the town of Selby was almost completely under water. Only the ancient abbey and the streets around the market place escaped inundation.

On 25 March the Lord Mayor of London opened a national fund 'for the relief of distress caused by the floods' to which the Government at once contributed £1 million. A huge number of organisations and individuals added their contributions. The States of Jersey contributed £5,000 and the American Red Cross sent 'fifty tons of soap, one thousand brushes and a thousand gallons of disinfectant'. Among individual donors, 'Mr E Cross of Aberdeen, who won £61,456 in a football pool, has given £500.' To no one's surprise, the month of March proved to be the wettest on record.

If the winter had at last ended, the Minister of Agriculture called the effects of the prolonged snowfall followed by the floods 'a disaster of the first magnitude', and the crisis was exacerbated by continuing fuel and food shortages in the aftermath of the war. This extraordinary year then saw one of the hottest and most prolonged drought summers on record. Temperatures climbed into the mid-thirties Celsius and apart from occasional ferocious thunderstorms triggered by the heat, there was no significant rain until well into autumn. Kent went fifty days without a drop of rainfall at one point, and October and November were the driest, sunniest and warmest on record in many parts of the country.

It was a golden summer for English cricket - Denis Compton and Bill Edrich both plundered well over 3,000 runs from the sun-baked pitches, with Compton's totals of 3,816 runs and eighteen centuries never likely to be equalled - but the drought did further damage to an already crippled harvest and led to severe food shortages and rising discontent among a war-weary population.

The most turbulent year of British weather ever recorded then concluded with one final spectacular: a thunderstorm on Christmas night that saw hailstones 'as big as marbles' fall across large parts of the South.

*Farmers struggled to fodder their sheep and millions were lost*

## Snow level with the roof.

When the "Great Snow" of 1947 began to fall, Rose Walmsley was living with her husband Fred and their two year-old daughter Heather at the isolated Newhouse Farm, 1,000 feet above sea level in Upper Wharfedale.

*'I can remember it as if it were yesterday. It hadn't been too bad a winter, and when it got through to February we thought we were over the worst of it. Then it started one Sunday. On the Friday we'd killed a pig and on the Saturday we salted it. On the Sunday, Fred was busy sawing wood, when suddenly the sky clouded over and the snowstorm came out of the north.*

*It was fine, powdery snow, whizzing round and swirling all over the place. It just kept on snowing and the terrific wind never stopped. By Tuesday the snow had blown level with the roof at the back of the house, forming a tunnel between the building and the drift. At the other side you couldn't even see the front door or windows.*

*I let Fred out of the back bedroom window into the snow tunnel using pig blocks - he was terrified I was going to drop him! He then forced his way round to the front and started to dig down to the outside door, making seven deep steps in the snow.*

*On the Wednesday things improved a bit and the sun came out. We knew we had to get Heather down to Grassington where she could safely stay with Fred's parents. All we had was a horse and a hay sled, so what we did was to line a dustbin with big hessian sacks. We then put Heather inside, tied the bin to the sled and Fred set off. I shall never forget the sight of this little figure with just her gloves visible as she clung on to the sides.*

*The snow had blown right over the tops of the gates and walls and it was freezing so hard that it was very crisp. It supported the weight of both the horse and the sled. Fred simply set off in a straightish line and eventually got down to Grassington. There were no tractors or snowploughs, so he picked up some meat and vegetables to leave at the hospital on the way back. They were running short and were right glad to see him. Finally he landed back; he had to leave the mare in the snow tunnel as there was nowhere else for it.*

*Soon afterwards it dulled in and started to snow again. It carried on like this for weeks on end, with two or three days of blizzard and then a bit of sunshine before more snow came. But there was never enough sun to start a thaw and so the drifts got higher and higher.*

*The wind never let up neither. It was what Fred called a lazy wind - it would blow through you rather than round you. It searched out every crack, forcing the snow through the sash windows. There was even a drift in the passageway formed by snow coming through the front door keyhole.*

*We had ten cows tied up inside the barn and each day it took us hours to muck out. Everything had to be carried by climbing up and over the snow. Then we had to find the trough so that we could bucket water to the cattle. Each night the drifts had blown over it. We also had to feed the horse which was still living inside the snow tunnel.*

*The biggest job of all was getting fuel inside the house. Fortunately we'd plenty of wood and we heaped the logs right up the chimney so that we had a roaring fire to keep us warm. We kept the boiler in the back kitchen full of water by opening the window and getting shovelfuls of snow, although it's amazing how much we needed.*

*It's a good thing we'd killed the pig just before the snow started as it meant we were all right for meat. The pork froze solid in the pantry and kept for weeks on end. We rolled the sides up for bacon and I cooked spare rib in all sorts of different ways. Often I would casserole it slowly in the fireside oven - it smelt wonderful and tasted delicious.*

*Fred always believed in keeping enough of everything to last a month. We had lots of oats and so had porridge every day. We also had plenty of spuds and were all right for turnips, which had been got in to to be chopped up for the cows. There was lots of flour in eight-stone bags, so Fred brought back a great block of yeast and I made bread. We also made our own butter and had eggs from our own hens, so we didn't do too bad considering.*

*It was April before the sun at last started to melt the snow and lambing time that year was dreadful. Even then it took ages for it all to go and there was plenty still around in June. Heather was away for eight or nine weeks before we dare bring her back.*

*We've had other bad winters but nothing like 1947. Today it's very different if it snows hard. Farms have electricity and tractors with a bucket up front. All we had were oil lamps and hand shovels.*

*It was a winter I shall never ever forget - I didn't see another woman from February to May.*

## PART III

# The Coldest Winter - 1962-63

If 1947 was the snowiest, the winter of 1962-63 was to prove the coldest, not just of the twentieth century but as far back as 1740. A cold snap beginning on 8 November 1962 ushered in one of the worst November blizzards ever recorded. Over the weekend of 16-17 November, gales gusting to eighty-five miles an hour hit the Scilly Isles and the sleet that covered the islands fell as snow over the British mainland, with heavy drifting in the fierce winds. Roads were blocked and traffic dislocated as far south as Somerset, Devon and Cornwall.

There was a brief respite, then on 3 December dense fog blanketed eighteen counties in England and Scotland. London Airport was closed, with freezing fog reducing runway visibility to fifty yards. Three Boeing jets from New York were diverted and all outgoing flights were grounded. Conditions were even worse in parts of London, with five yard visibility reported from Wimbledon and Malden, and "nil" visibility in the Wembley area.

The following day the fog persisted with a toxic cloud of "smog" - a mixture of smoke from coal fires and sulphur dioxide - returning to the streets of London and Britain's other big cities, the worst since the 'Great Smog' of 1952. Concentrations of smoke and sulphur dioxide were already nine and five times the normal average respectively. Four lives were lost in crashes in the fog. The worst incidents were multiple pile-ups on fogbound motorways, but buses and trains were also involved in collisions. Thousands of cars were stuck in jams on the North Circular Road, many drivers abandoned their vehicles, 'many other halted behind, unaware that they had done so'. The Duchess of Kent, with her five month old son, was due to fly from Stansted to Hong Kong, but got no further than the outskirts of North London before abandoning the attempt and returning ot Kensington Palace.

The only aircraft to land at London Airport all day was 'a twin-engined Ministry of Aviation Varsity Aircraft from the Blind Landing Experimental Unit at Bedford', seizing the chance to test their equipment. BEA alone cancelled 82 flights from Heathrow and though at one time more than 1,000 frustrated would-be passengers crushed into 'final departure lounges', all were turned back. Hundreds of rush-hour passengers on the 5.06 pm train from Liverpool Street to Chingford 'had to get out and walk' when it derailed at St James's Street station, Walthamstow. A farmer from Islip near Oxford was killed when his tractor was hit by a diesel train on a fogbound level-crossing and twelve people were injured in Bradford when

*The snow rurned the normally bustling centre of Bradford, Yorkshre, into a ghost town.*

a car left the road in the fog and ran into a bus queue. Thirteen people were hospitalised after two London Transport buses collided at Manor Park, and a woman drowned in the Rochdale Canal at Failsworth, Manchester, after 'missing her way while crossing a bridge'.

By the next day, 5 December, the Air Ministry was reporting that smoke and sulphur pollution was 'as great as in the Great Smog of 1952'. Ford's at Dagenham sent their workforce home at 2pm that day, though forty percent of the dayshift had failed to turn up for work anyway. Shipping in the Port of London was at a standstill, the Clyde was also closed to shipping and London Airport was again closed. Fans of Tottenham Hotspurs travelled to Glasgow for the European Cup Winners Cup tie against Rangers found when they arrived that the game had been called off because of the fog. Bus services were suspended and London Transport ordered drivers to make for the nearest garage at once. Scotland Yard announced that there had been twenty-eight sudden deaths in London that day, 'the usual figure is six to eight a day'. There were other deaths caused by the fog. A fireman whose goods train had developed an engine failure at Kentish Town West plunged forty feet to his death from a viaduct as he walked up the line to lay detonators on the track as a warning to other trains.

On the following day, Thursday 6 December, it was reported that cases of pneumonia had trebled in Glasgow since the smog began. Thirty people, mostly elderly, had been admitted to St James's Hospital in Leeds, suffering from chest complaints and Mr R Dalley, Leeds City Analyst, said that sulphur dioxide levels in the atmosphere were now the highest ever recorded, exceeding those in London during the Great Smog. Concentrations of smoke were now ten times, and sulphur dioxide fourteen times normal levels. 'There is no doubt that without the Clean Air Act of 1956, the conditions would have been much worse than 1952', Mr Dalley said. A further thirty sudden deaths were reported to London's Metropolitan Police that day, taking the total since midnight on Monday to ninety.

All forms of transport were seriously disrupted and thousands of drivers simply abandoned their cars and walked. One Bradford man drove to a party hanging out of the window as he tried to see the white line in the middle of the road, while his passenger leant out of the other side trying to spot the kerb. When they eventually reached their destination, after taking forty-five minutes to drive three miles - longer than it would have taken to walk - they discovered that the party had been cancelled because of the fog.

At times visibility over the whole of the Greater London area was less than five yards, and several leading London stores complained that the fog was keeping Christmas shoppers away, with turnover well down. London Airport remained closed, and chaos on road and rail continued. All London buses were again halted for the night at seven that evening. A railway shunter was killed and three workers injured when two empty trains collided at Gillingham and half the passengers on the twelve-coach 8am train from London to Brighton were stranded at Streatham when the train 'broke in two and the second six coaches slowed down and stopped without the passengers realising that they had been left behind.' The loose carriages were eventually pushed into Streatham Common station by the next London to Brighton train.

On the morning of Friday 7 December, a Middle East Airlines Comet taxied up and down the main runway at London Airport with the pilot running the engines 'at high power'. The heat from the engines cleared a path through the fog for a few minutes, during which the Comet took off, bound for Rome and Beirut. It was the first aircraft to take off at Heathrow in four days. A plane carrying the 79 year old Princess Alice also took off successfully from Bovingdon, Hertfordshire, in ninety yard visibility but was then forced to turn back by fog over Holland. By then the fog had largely cleared over London Airport and she landed safely and then travelled to Kensington Palace by car. She considered her options over a spot of lunch and then set off again, travelling by car and boat-train from Harwich to the Hook of Holland. The four days of fog at London Airport had caused the cancellation of 300 BEA services at a cost of £150,000.

What was to prove to be the last of the old-style smogs was over, but 'the appearance of cars left standing unprotected during the past few days is sufficient proof that curtains and the like will have collected their fair proportion of London's combustion products - including acid, which is better washed out than left in.' The Times of 8 December editorialised that '"pea-soupers" and "London Particulars" were undoubtedly worse cursed by blackness, but they were not any worse - and some would say they were better - in respect to lethal fumes.' Dr JA Scott, the London County Council's Medical Officer, later estimated that the capital's total deaths from respiratory problems caused by the four-day smog had been 340 people - other estimates ranged as high as 1,000 - and The Times went on to call for action to remove toxins from the air, claiming that smokeless zones were 'in a sense confidence tricks', leaving air apparently cleaner but full of toxins like sulphur dioxide. That drew a furious response from the Society for Clean Air, which deplored 'hysterical, hasty and ill-considered press statements'.

*Only foot-traffic is moving as a long line of stranded vehicles wait for some sign of gritters.*

After that unpleasant prelude, the winter of 1962-63 began in earnest at Christmas. On Christmas Eve more fog and black ice caused thirty-five accidents in Leicestershire alone; two people were killed and more than twenty injured. The newspapers reported a record Christmas 'rush to the sun from London Airport, with 60,000 departures since last Thursday'. They left just in time...

A belt of rain over northern Scotland turned to snow as it moved south, giving Glasgow its first white Christmas since 1938 and Guernsey its first since 1919. Temperatures plummeted. Kew had its coldest Christmas since 1944, Gatwick, Hurn, Abingdon, Ross-on-Wye and Cardiff all recorded -9°C in the early hours of Boxing Day and in Birmingham dropped to -11°C. Ground temperatures were even lower; it was -15° C at Birmingham, and

*A motorist making the seemingly futile attempt to extricate his car from a queue of snowed-up and abandoned vehicles.*

snow then falling on already frozen road surfaces caused traffic chaos, with the AA reporting that all roads from the Midlands to southern Scotland were 'very dangerous'. Not to be outdone, the RAC added that 'Drivers are slipping about as if they were learners on skates.' By sunset on Boxing Day heavy snow was falling all along the South Coast of England and even in the Scilly Isles.

By the following day, 27 December, the North was 'gripped by Arctic conditions', there was two inches of snow in the Channel Islands and it was a foot deep across much of southern England. Roads over the Mendips were closed and the London to Hastings road had 'deep drifts in several places' and, said the RAC, cars were sliding off the roads 'like spinning tops'. Letters were scattered all over a Lowestoft pavement when a bus skidded off an icy road into a post-box. Many flights were cancelled - Gatwick had almost a foot of snow on the runways - and rail delays made thousands of Londoners late as they returned to work after the Christmas break. Eire was also badly affected with Dublin roads 'hazardous'.

A fresh blizzard over South-West England and South Wales on 29 and 30 December produced snowdrifts up to twenty feet deep. There was heavy snowfall in the North and Scotland too, with the RAC reporting 'deteriorating road conditions'. Villages were cut off, roads and railways blocked, telephone wires and power cables brought down. Thousands of sheep, ponies and cattle were buried under the snow.

Seventy-one passengers from two coaches were trapped overnight at the Clay Pigeon Cafe at Warden Hill near Evershot, Yeovil. Among the passengers were a dozen children and a month-old baby girl, Tracie Gallagher, from Birmingham. A local farmer provided her mother with a cardboard box as a crib, where the child, wrapped in blankets, spent the night. They were still trapped twenty-four hours later, though 'a man on skis' had got through with some baby food. A snow plough was reported to be on its way, but the cafe owner said there was 'little hope' of it being able to reach them that day. Another two parties of passengers stranded at Lee Mill near Plymouth, were rescued by police in mid-morning after a frozen night in their coaches. Yet another pair of buses, carrying twenty-five passengers, were stranded by twenty-foot drifts on the Blandford to Salisbury Road in Dorset before eventually being rescued by a farmer 'with a tractor and horsebox'. Farmer Mr Charles Burt, his wife and their four small children, spent all night in their car after becoming stuck in a snowdrift on the Isle of Wight. After eleven hours 'huddled in the back', they were found by council workers, who dug them out.

A baby boy was born in a snowbound ambulance at Praze on the Helston to Redruth road in Cornwall, and a helicopter dropped yeast and milk to the bakery at Dartmoor Prison where supplies were running perilously low. Seventy railway enthusiasts had their enthusiasm tested to the limits after being trapped on a train at Tavistock all Saturday night. Police took food to them on Sunday and eventually they struggled through the snow on foot to an emergency rest centre set up by the Women's Voluntary Service. The night express from Penzance to London Paddington left on time at 8.50pm but 'had such a struggle through Cornwall' that it didn't leave Plymouth until five o'clock the following morning. Frozen points then delayed it further at Bristol and it eventually arrived six hours fifty minutes late. Troops were called in to clear snow from the sidings at Willesden to release rolling stock for the Monday morning rush hour, and many Southern Region lines were blocked, with Orpington closed until Sunday afternoon.

An ambulance on the way to pick up an expectant mother at Farningham, Kent, was stuck

*A stranded "half-timbered" Morris 1000 Traveller on a road near Bournemouth.*

in several feet of snow for nearly six hours before two snowploughs managed to free it. An RAF helicopter rescued four people trapped all night in a car at Blackmore Gate on Exmoor and a young mother-to-be, Angela Christmas, was rescued unconscious but alive after being trapped all night with her husband in a six-foot drift at Crockenhill, Kent. PC Peacock of Swanley Police rescued them as he was returning home from night duty.

However, several lives were lost because of the snow. A sixty-five year-old railway worker, John Warren of South Benfleet, was killed by a train as he tried to clear snow and ice from frozen points, a sixty year-old milk roundsman, William Starkey, collapsed and died while making deliveries in the snow, and a husband and wife, Arthur and Daisy Barber, were found dead in their car after being trapped by drifts on Osmington Hill, Weymouth. Their daughter, Mrs Sheila Reed, her seven year-old son, Ian, and Thomas Cumbes of Dorchester, who were trapped with them, all survived.

On New Year's Eve, the one month-old baby, Tracie Gallagher, and her mother were among sixteen passengers - the others all elderly or sick - rescued from the Clay Pigeon Cafe by a Royal Naval helicopter flying from Portland in Dorset. All the other passengers were left behind, but were being rescued ten at a time by a farmer's tractor and trailer, running 'a shuttle service' to Cattistock, from where coaches took them to Dorchester. At Denton Mains, near Gilsland in Cumberland, twenty people were trapped for fourteen hours in two service buses. They were only 500 yards from a farm, but in the blinding snow, no one realised that, and when farmer Thomas Graham eventually discovered them, 'Everyone in the bus looked ill'.

Troops were used in many areas to open roads and rescue stranded people. Soldiers from the Wessex Brigade depot at Honiton and the RASC at Yeovil were brought up in a special train and worked through the night in an effort to clear the blocked railway line between Tavistock and Okehampton. Men of the Gurkha Regiment - who presumably knew a thing or two about snow - cut their way through 'a big snowdrift' near Amesbury, Wilts, where 'sick and old people were marooned'. Sappers from Chatham used bulldozers to clear roads in Kent and Sussex, and on Salisbury Plain an Army Operations Room was set up to co-ordinate rescue missions to isolated farms. Despite the best efforts, Weymouth, Bridport and many other areas were still completely cut off by 'high drifts, icy floods and abandoned vehicles'. As night fell on this, the last day of 1962, it was still snowing heavily and almost the whole of Britain was covered by snow.

It was not going to be a happy new year for farmers. Farm dairies were 'overflowing for lack of churns' and farmers had to pour thousands of gallons of uncollected milk down the drain. In Sussex, ten thousand gallons of milk was standing at the roadside, waiting in vain for tankers to collect it, while the Milk Marketing Board had begun making hardship payments from an emergency fund. While there was a glut of milk on the farms, doorstep deliveries in London and the Home Counties were being threatened by a shortage of milk and of bottles to put it in. A spokesman for National Dairies warned 'unless customers make efforts to hand in empties, there could have to be reduced supplies.' London's milk supplies were given a temporary boost by 'a switch of thousands of gallons from Cheshire, the Midlands and East Anglia', but the position remained critical.

Farmers in the Fens and East Anglia who had 'already lost thousands of pounds worth of celery because of the frost', were fighting to save 175,000 tons of sugar beet, valued at £1 million, which was frozen into the ground. The Cantley sugar beet factory 'faced closure' if

*Commuters inching over the Clifton Suspension Bridge in Bristol.*

supplies of beets did not arrive soon. Tens of thousands of animals were buried by snowdrifts. Thousands of sheep and more than two thousand ponies were trapped on Dartmoor and the Horses and Ponies Protection Association issued an appeal 'for the rescue of the animals'. Whipsnade Zoo was cut off and keepers struggled on foot through twelve-foot drifts to feed the animals. Human food supplies were also threatened. A spokesman for the National Federation of Fruit and Potato Traders, perhaps shedding a few crocodile tears, announced that prices of vegetables and many fruits had 'shot up' since the snow began and would reach a peak later that week.

At London Airport, two runways had now been cleared, but seventeen BEA Comets, Vanguards and Viscounts were still grounded by three-foot drifts, more than twenty-four hours after the last snow had fallen there. A Ministry of Aviation spokesman claimed 'We are clearing the taxi-ways and terminal aprons as fast as we possibly can,' but airline chiefs were furious at the delays. Meanwhile a spokesman for British Railways Western Region was praising its workers who had walked to work 'many trudging distances of from three to eight

*Edinburgh to London sleeping car train snowed up near Dent station.*

miles. Two teenagers in the Enquiry Office at Bristol stayed on duty from 9am to 10.30 pm on Sunday, and then walked home. A snow plough crew remained on duty for eighteen hours.'

The gales that were still piling up drifts on the high ground, also brought further problems off the coast. Heavy seas backed by the fierce winds did 'widespread damage' at Paignton in Devon. 'A thirty-two-foot naval cutter was dashed against the West Quay and sank. Small boats were damaged and sea-front kiosks blown over.' The motorship Egton broke her moorings in the River Wear at Sunderland. Two helicopters of 22 (Search & Rescue) Squadron at RAF Chivenor, Devon, spent ten and a half hours - from first light until after dark - flying rescue missions and food drops to isolated areas. During four sorties to take food to Lynton, 'the wind was so strong that the helicopters were 'flying sideways like crabs'. Gales also caused severe damage on land, leaving 'a trail of destruction over the Macclesfield area'. Only one route from Macclesfield to Bollington remained open as 'trees fell like ninepins'. The winds also toppled the 200-foot mill chimney of Firgrove Mill in Rochdale. It crashed across the Rochdale Canal and tore a hole through the wall of J & J Makin's paper mill, 'where the night-shift was working. Two workmen ran for their lives.'

A road was closed in Royton, Lancashire, after 'a large corrugated iron engineering workshop' collapsed, and 300 people were evacuated from the Ritz Cinema in Stockport after 'large roof tiles' were torn off by the wind. A chimney stack tore a twenty-foot hole in the roof of a wines and spirits firm in Chesterfield, gable ends collapsed in several towns, and shop windows were blown in. Twelve Euston to Glasgow expresses were delayed after 'the stay-wires' of a telegraph pole snapped in the gales and it 'swayed over the line'. An Air Ministry spokesman helpfully explained that 'in this type of easterly airstream, funnel effects caused by local land contours can produce these freak winds'.

Water from a burst pipe seeped through three floors of 'a large furnishing store in Hammersmith' causing hundreds of thousands of pounds damage, and there were further human tragedies and near-tragedies. Roadmen found Mrs Donna Douglas, 'Cinderella in the pantomime at the Empire Theatre, Sunderland', and Mr Charles William Brown of Railway Hotel, Miller's Dale, Derbyshire, unconscious in a car buried by snow at Carter Bar on the A68 at Jedburgh. The car engine was still running and 'the occupants overcome by fumes'. Both had to be given oxygen and were very fortunate to survive. A sixty year-old woman, Mrs Wanda Zakrzewska of Shalbourne near Marlborough in Wiltshire, was less lucky. She had taken her dog for a walk in the snow and was later found dead 'in a trench near her home'. Her unharmed dog was still sitting next to her body. A railway signalman, fifty-five year-old Albert Heddington, collapsed and died as he trudged through the snow to work at Greenfield goods yard near Oldham in Lancashire. And George Rose, forty-six, of Tunbridge Wells, also collapsed and died while shovelling snow from the steps of the Rehoboth Baptist Chapel.

The new year had begun as the old one had ended, with heavy snow, described by Gordon Manley, Professor of Geography at Bedford College, London University, as 'the worst over southern England for eighty-two years.' In the hundred years for which records were available, he said, the only worse year was 1881, when on January 18, 'there was a drift fifteen foot deep in Oxford Circus' in the heart of London. More than 200 classified roads and thousands of minor roads were 'strangled by slush, snow and ice' according to the RAC, despite the efforts of 'more than 1,000 snow-clearing appliances' in England and Wales and, said an AA spokesman, 'the only thing going up the M1 is snow'.

Two inches of snow fell 'in little more than an hour' in Hendon and by midnight it was snowing hard from Kent to Somerset.

In Marylebone, 'snowploughs and mechanical diggers will come into action again, and the movement of traffic, let alone the parking of cars, will be gravely impeded by banks of snow and ruts.' In plutocratic Kensington 'they make no use of snowploughs, but shovel the snow away by hand. Some of it follows the route of the Borough's refuse to a wharf in Chelsea, some of it is run down the large sewers of Kensington'. As if drivers did not face enough problems it was reported that stockpiles of salt for gritting roads were running out. Westminster alone had already used 1,100 tons, and Wandsworth had used its entire 700 ton stock and been told that fresh supplies would not arrive for three or four days. Westminster had been given the same response with the added proviso 'weather permitting'.

At the Winsford salt mine in Cheshire, source of virtually all the domestic output of salt for roads, lorries were queuing for two hours before being loaded. British Railways, 'acknowledging that their transport of salt is not at the moment imbued with any sense of urgency, can point out that their entire service is slowed down by the snow which has created the demand for salt'. ICI, owner of the Winsford mine and the chief supplier of road salt, claimed to have done all it could to make supplies available. 'It is chiefly through their persuasive advertising that councils got maximum stocks of salt at "lowest summer prices" months ago.'

To add to the misery, it was reported that meetings of power workers at Barking and Tilbury had voted to join the unofficial work-to-rule and overtime ban already being operated at twelve other London power stations. Power-cuts meant that the supply of car bodies from the Pressed Steel Company's Swindon plant was reduced 'to a trickle' and production of the Morris 100 at Cowley, Oxford, stopped at lunchtime. Production of the A60 at the Austin factory in Birmingham 'resumed again this morning, but stopped at lunchtime' and 500 workers were laid off.

The next day, Thursday 3 January, the work-to-rule led to further power-cuts in London and Essex and part of Barking power station had to be closed altogether. Supplies in Streatham, Hindhead, Hadley in Surrey and Liphook in Hampshire were also cut. Fifteen hours of negotiations between the unions and the Electricity Council over the pay claim for the 128,000 manual workers in the industry broke down without agreement that night. The Electricity Council's increased offer of twopence-halfpenny an hour (one new pence), just over a halfpenny more than their previous offer, was rejected by the unions.

There was more depressing news for householders with the announcement that 'the milk situation' was 'threatening to be desperate' as emergency arrangements for transporting milk from the South-West to London had, like the power workers negotiations, 'broken down'. Supplies of vegetables were 'very short' and wholesale potato prices on the London markets had risen to '£24-£28 a ton for whites and £25 - £32 for King Edwards'. In addition, the 'condition of some loaded from field clamps is reputed to be far from good'. Agriculture Minister Nicholas Soames chose that moment to announce that if Britain joined the European Common Market, 'some upward adjustments' in food prices were inevitable. The British Sugar Corporation estimated that ten per cent of the previous year's sugar beet crop of 500,000 tons had still not been lifted from the ground and 'would soon be a write-off. Where there has been much frost [ie everywhere] the condition of the beets is still doubtful and some may well already be too far gone to process.'

*Blizzards and icy roads made many city roads impassable.*

*Snowdrifts burying the face of a building meet the icicles hanging from the roof*

If the fate of the sugar beet crop did not disturb too many suburban slumbers, the 'thousands of dustbins overflowing' because the refuse collectors had been unable to get through, were becoming a serious problem. In 'hilly Hampstead' the council was collecting fifty less tons of refuse a day than normal, Wandsworth, concentrating on trade waste, distributed paper sacks... which snow and slush immediately rendered useless.

'Snow and sheet ice' had now combined to keep the South-West of England and South Wales paralysed. There was 'heavy snow' in Swansea and blizzards blowing across Devon cut off many more villages there, and grounded helicopters from RAF Chivenor and RAF Culdrose in Cornwall. The AA reported that 'conditions in many areas are as bad, if not worse, than at any time since the snow onslaught began a week ago'.

All roads in and out of Bristol were blocked, 4,000 workers at Morris Motors in Cowley, Oxford, were sent home early, partly because of a shortage of components and partly to help them reach their homes before nightfall, and more than a thousand workers were unable to reach the Atomic Research Centre at Harwell. United States airmen had to take to the fields in heavy vehicles to take supplies to the villages of Leckhampstead and Chaddleworth in Berkshire. In the North, all the trans-Pennine roads were impassable because of drifts. In Monmouthshire a retired haulage contractor, Wilfred Benfield, 59, 'had only just taken up his shovel' to help a neighbour clear snow when he collapsed and died. A steelworker, Edward Glyndwr Davies, 54, was found dead. lying face-downwards in the snow in the Goytre Valley in Port Talbot, and Peter Henretty, 50, a lorry driver of Edinburgh collapsed and died as he exchanged names and addresses after 'a minor accident' on a snowbound section of the Great North Road near Berwick'. The Fire Protection Agency warned of the dangers of an increased death toll because 'unthinking householders' were using used 'blow-lamps, lighted candles and a variety of oil heaters to thaw out frozen pipes, often with disastrous results.'

*Milk deliveries froze solid on the doorstep, with the bottle-top pushed up an inch or more on a column of frozen milk.*

There was a startling variety of weather across the country, with just one thing in common - all of it was bad. As well as the continuing blizzards in the West, there was fog in Kent, and widespread snow and 'icy rain' in the North, while Gatwick Airport was brought to a halt by freezing snow and slush, causing cancellations and overnight delays of flights to France, Germany, Switzerland and the Channel Islands.

There was a slight thaw in the South on 6 January but that only brought fresh problems as the AA warned that roads were 'already cracked or potholed' by the frost and snow, with some potholes 'over six inches deep. This can easily happen after a prolonged spell of frost or snow and motorists should be on their guard.' Despite the brief warming, the AA added that 'snow- and ice-belts' gripped the South-West, and the South

Coast from Devon to Hampshire and northwards to Oxfordshire and Gloucestershire. Almost the whole of Wales, Shropshire and Staffordshire was also snowbound and everywhere 'east of a line from Scotch Corner in Yorkshire to Gairloch in North-West Scotland'. Only Cheshire and Lincolnshire were reported by the AA to be free of the 'Alpine weather conditions.' One of the most notorious winter black spots, the A68 around Carter Bar, was now blocked for twelve miles. An avalanche of hundreds of tons of rock and snow was blocking almost a quarter of a mile of the mainline between Edinburgh and Carlisle near Galashiels, and foghorns had to be used to warn shipping of danger after further blizzards reduced visibility at Flamborough Head in Yorkshire to nil.

There was better news for some trapped by the snow. A shepherd, J Smith, and his wife snowbound for sixteen days at the top of the Lammermuir Hills in Scotland had been reached with supplies, and holidaymakers at the Forest Inn at Hexworthy on Dartmoor - snowbound there since Christmas Day - were finally rescued by the driver of a Landrover who 'broke through to them'. A Lakeland terrier trapped for sixty hours at Hardcastle Crags near Hebden Bridge scrambled free just as rescuers were giving up for the night. It leapt into the arms of its owner, David Dalgliesh, 'the head keeper on Lord Savile's 16,000 acre estate, and was taken home for its first meal in three days'. There was more good news for animal lovers when police reported that 1,500 stranded sheep, cattle and ponies on Dartmoor had now been rescued and foddered. The rescuers had encountered drifts between twenty and thirty feet deep, and many thousands more animals were still buried under the snow. And a disconsolate Secretary of the National Mouse Club, Percy Ashley, reported that the freezing temperatures had led to 'a sharp fall in the birth rate among Britain's prize mice'. Mr Ashley warned that 'Entries in the baby mice classes at the big January shows will be hit.'

More snow fell on 7 January, heaviest over Scotland, and road conditions everywhere were described by the AA's hyperbolic spokesman as 'satanical'. A farmer at Edmundbyers in County Durham said the snow was 'the worst I have seen. Some of the drifts are as high as houses'. Diesel froze in buses and lorries at Grantown-on-Spey in Morayshire, where the temperature fell to -21C, and railway workers were using dynamite in an attempt to clear the blocked railway line at Riccarton between Carlisle and Edinburgh. On Dartmoor sheep buried by snowdrifts were being 'eaten alive by hungry foxes which are burrowing through the snow to get to them' and police in the New Forest area issued a warning that 'hunger-crazed ponies' might attack people carrying food. A hotel worker, Mrs Linda Parker, had already been 'set upon by two black mares in foal'. As she walked home to her caravan near Brocklehurst, the ponies 'ran up behind her, bit her and kicked her. She was rescued by two men who chased off the ponies with sticks.' A corporation workman got some revenge against the animal kingdom for such indignities while unloading snow from a lorry parked on a slipway near Margate in Kent. He dropped his shovel into the water and as he bent down to pick it up he saw 'a ten pound cod, measuring two foot six inches' in the water. 'I scooped up my shovel quickly and the cod came with it. It was alive and kicking until I hit it with the shovel'.

Over the next few days the vicious cold continued. There were fresh snowfalls in the North and Scotland on 11 January and by the next day "ice-floes" were forming in the River Thames at Windsor. A bystander saw 'a solitary skiff being rowed along the river, noisily hitting the chunks of ice'. The Thames was soon frozen right across. 'For the first time in memory', the River Urun 'the second fastest-flowing river in Britain' was frozen over completely where it met the sea at Littlehampton in Sussex. People skated on the ice, fourteen inches thick, that covered the River Avon at Tewkesbury, and at Newnham-on-Severn, three men 'lashed themselves together and walked across the river's icy surface'. Many other lakes and rivers froze, ice covered harbours, the sea froze for some distance from the Kent shore and there were ice-floes and pack-ice off the coasts. Huge blocks of ice were formed on beaches as the spray from breaking waves froze as it landed, and at Torquay and Herne Bay in Kent, 'sea water froze for the first time in living memory, as it crashed over 100 yards of promenade'. The fountains in Trafalgar Square in London were also frozen solid and the pigeons picked their way across the sheet ice surrounding them.

The misery of the bitter cold was compounded by the continuation of the work-to-rule by power station workers. While union officials continued to debate whether to make the

*Workers digging through a twelve foot drift to free a stranded lorry laden with cattle feed*

dispute official, leaders of the unofficial action called for 'a more rigid application' of the existing work-to-rule. There were voltage reductions at each peak period from 'the Wash to the Sussex Coast'. Fires and lights were dimmer and signalling and switching on Southern Region railway lines was also affected. A surgeon, Frank Musgrove, declared Britain was 'like a jungle' after having to work by candlelight to save a mother whose baby died during a power-cut at Wanstead Hospital, Essex. Priority in the distribution of coal was given to steam trains carrying supplies of food, coal, oil and petrol. As the power workers' action escalated still further, electricity cuts darkened streets, switched off traffic lights, closed cinemas and theatres, and left many other businesses operating by candlelight. Old paraffin heaters were pressed back into service and many fires were caused through heaters being knocked over, or left unattended, often in lofts, in an effort to defrost frozen pipes. Many drivers left a small paraffin heater burning all night under the engine block of their car when they had parked it in their garage. Although such heaters were normally fitted with a safety gauze, it was a miracle that the lethal combination of a naked flame and vapour from petrol engines did not produce more fires and explosions.

Water pipes and mains burst in the cold. It was reported that 200,000 gallons of water a day were being lost to leaks and bursts in the Windsor area alone, and tarpaulins filled with insulation material were hung from Windsor Bridge in an attempt to protect the water-main running along the underside of the bridge, and Brighton Corporation wrapped emergency standpipes in sheets of polythene, 'Inside is a red lamp to prevent the standpipe from freezing.' The Windsor Express reported that two children aged five and six had turned up clutching 'ready-pasted' toothbrushes at their dentist's surgery, asking if they could clean their teeth there as the pipes at home were frozen. Pipes freezing and then bursting as they thawed also let thousands of children have days off school and gave plumbers one of their biggest bonanzas in years. As further proof of the truth of the old adage about ill winds, a trader gleefully reported a 'fantastic demand' for buckets.

On 16 January snow returned to the UK with significant falls continuing over several hours in most parts of the country apart from the South-West, which was probably due a reprieve. There was a further very harsh frost that night - the AA described Britain as 'a vast ice patch' - and in London the temperature had now not risen above 3°C for the twenty-five consecutive days since the cold spell began on 23 December. It was another new milestone of misery, the longest comparable spell in the past had been twenty-four days in 1890 and 1895.

Bus passengers in Swansea circulated a petition urging the installation of heaters on the buses, but not all passengers signed it; 'some said they could not sign as their hands were too cold to hold the pen.' There was as different cause of "bus rage" in Doncaster, where a fourteen year-old schoolboy, James Boyle, was turfed off a bus and left to walk eight miles home through thick snow because he'd lost his bus-fare. When he told the conductor, she said, 'Well, you'll have to get off then, won't you?' But 100 yards down the road he was allowed to reboard the bus, after one of the other passengers paid his fare for him. When asked why the boy could not simply have given his name and address to the conductor, the Yorkshire Traction Company's spokesman, Mr C Dean, replied, 'If I walk into a shop and find I have no money, I don't leave my name and address. It may cause hardship, but unless there is someone willing to pay the fare, that is the way it is.'

*The bananas must get through... With no gritters on the streets, drivers had to take matters into their own hands. This lorry driver has wheelbarrowed some grit from a nearby stockpile to try and get his lorry up a snow-covered hill.*

Water and gas mains were again affected by the frost and there was a tragedy in Salford, Lancashire, when five people died of gas poisoning after a four-inch gas main fractured only seven or eight feet from the victims' house. The sole survivor, Ann Hardy, who was just eleven years old, lost her parents, her grandmother and her two brothers. Two days after the incident, doctors had still not broken the news to her that she was now an orphan. She was said to be 'a little better' but doctors advised that she should not be told 'until she is well again, when a priest may be asked to break the news'. Another awful tragedy was averted by inches, thanks to the quick reactions of a lorry driver. Arthur Baxter braked his eleven-ton oil tanker to an emergency stop at Caenby Corner roundabout near Lincoln, when 'two bundles' fell from the car in front of him as it took the roundabout. He got out and found two brothers aged nine and eleven lying in the road. 'My stomach turned over,' he said, 'when I saw the children only about two inches from my wheels. They were screaming and so was a lady in the car'. They survived, having suffered only 'bruising and shaking'.

*28 February 1963: Teams of men clearing the snow from Cheltenham racecourse in preparation for the National Hunt Festival beginning on 12 March 1963. The entire course had to be cleared of snow by hand, since machines would have damaged the turf.*

A tragedy of a somewhat less serious nature was brewing in Doncaster, where a 'women's underwear crisis' was giving the town's female inhabitants some anxious moments. The manageress of a women's outfitters reported a dearth of warm underwear saying 'I have had to turn away dozens of orders. The makers say they have been taken unawares by the severity of the weather.' But 'London hosiery manufacturers who put on their "Beatniks" - knee-length, woolly bloomers - as a gimmick and found it an all-time winner' were besieged with orders. A spokesman for the company made the curious statement that 'Girls bought the garments for a laugh at first, but find in the cold weather that they can't get out of them.' Meanwhile a looming crisis over the untidy gardens of council homes at Norton, near Malton, was averted when council officials, recognising the impossibility of gardening in sub-zero temperatures gave tenants an extra month to put their unruly gardens in order.

There was an echo of 1947 on 19 January, when, just as in that year, newspapers published a picture of a farmer (this time at Waldringfield in Suffolk), using a pneumatic drill to harvest his carrots and parsnips from his rock-hard frozen fields. 'A production rate of a ton a day is claimed, but some of the carrots are damaged.' They had to be broken out of the clods of earth with a hammer.

That day there were further two hour power-cuts in London. Traffic lights failed in the West End and extra police were called in to deal with the resulting traffic chaos. AA emergency staff dealing with breakdown calls had to work by candlelight. West End cinemas were blacked out, films stopping in mid-screening and the start of the Kurt Weill Opera at Sadler's Wells was delayed by forty-five minutes until the power came back on.

The temperatures remained bitterly low and the easterly gales, feeling as if they had blown all the way from Siberia, made the situation even worse. Salting the roads was becoming ineffective - the temperatures were so low that even salt water froze. One of the heaters from London Airport had to be installed at the Corporation Depot in Alma Road, Windsor, because the piles of rock salt were frozen into boulders that were impossible to load and use on the gritting lorries. The Western Region of British Rail cancelled several trains because of icing in the savage frost. Steam and diesel locomotives were 'running with difficulty' and 'some were even freezing up while running'.

The vicious easterly gales brought fresh blizzards with them and 'Britain was almost cut in half' with every cross-border road between England and Scotland blocked. More than 200 vehicles were stranded on the A66 at Stainmore - like Carter Bar, a notorious winter black spot - after 'the worst blizzard in sixteen years'. Some drivers abandoned their cars and fought their way through the drifts 'to spend the night at a moorland cafe'. Forty-eight soldiers missing after a twenty-four hour exercise on Dartmoor were found in an empty cottage south of Princetown. All were suffering from frostbite. Their boots had frozen solid when they took them off to bivouac at night and, unable to put them back on again, they

*Bad news for parents, good news for kids - Scottish schoolchildren making good use of a fresh snowfall*

*A not very sun-blessed Morris 1000 in the winter of 1962-63*

walked 'in stockinged feet' 200 yards to the cottage where they were found by a search party. Police patrol cars were sent to search for the Archbishop of Canterbury after he was reported 'very late'. He arrived safely at Canterbury at 10pm after a six hour journey from Lambeth Palace, most of which time had been spent in 'a huge traffic jam near Swanley'. On the railways, gangs of men battled for hours to dig out an Edinburgh to London express that was completely covered by a snowdrift 'on a wild Pennine moor' near Dent in North Yorkshire. The railwaymen eventually managed to release the rear three coaches and the fifty-seven passengers were taken the sixty miles back to Carlisle and 'given refreshments' before continuing their journey via Newcastle. 100 railwaymen were meanwhile continuing to try and free the locomotive and the first five coaches of the snowbound train. Other lines in the area were blocked by fallen trees brought down by gales.

The following day, in 'one of the biggest airlifts ever, four jet-powered helicopters' evacuated 283 staff from the Fylingdales Early Warning Station on the North Yorkshire Moors, who had been trapped there for five days with fuel supplies running low. 'Almost every piece of snow-clearing equipment had been lost' in an attempt to force a way through twelve-foot drifts and Max Goldman, the American site manager of the euphemistically named Radio Corporation of America at Fylingdales, said that weather conditions were 'unbelievable', worse than anything he had experienced during the construction of an early warning station in Alaska. 'The wind is blowing at between seventy and eight miles an hour,' he said, 'and no one is allowed to go outside alone. It is too dangerous out there.' The snow was being driven so hard by the gale that it covered a double-decker bus in less than an hour. Despite the gales, the staff were ferried out in batches of eight to ten, and taken to Whitby twelve miles away. 111 civilian workers were still stranded at Fylingdales when flights were suspended at nightfall, and they were forced to endure another night trapped on the moors.

The gales and snow sweeping the country created chaos everywhere. Despite the police putting 'every available man on point duty' there was a fifteen mile traffic-jam in Halifax, as thousands of vehicles made for the only road still open into Lancashire, and at one stage it was taking five hours to cover the eleven miles from Halifax to Todmorden. Workmen gave up any hope of opening any of the other trans-Pennine roads; 'as fast as a stretch of road is cleared, it fills in again as the snow is whipped up by the strong winds.' On the worst stretches of road, drifts of thirty and forty feet deep were reported. On the Settle-Carlisle line, railwaymen worked all day for the second day running in an unavailing attempt to free the locomotive and five coaches trapped at Dent. Flights were halted yet again at London Airport after an Austrian airliner hit a patch of 'glazed ice' and skidded off the runway. BEA had to cancel forty-four flights and divert a dozen others. The AA said that roads in eighty of the eighty-six English, Welsh and Scottish counties were now impassable because of snow or abandoned vehicles. British Rail's Western Region announced that it was cancelling twenty-three services a day 'until the weather improves' and there were heavy delays on the Southern Region because of snow, freezing rain and icing of rails and points. The 10.10am from Ramsgate did not reach Victoria until 5.20pm.

Two men were missing after their tug capsized in the icy waters of the River Humber. The sixty-five year-old skipper of the tug, Mr H Ellerby, and the Engineer, thirty year-old Benjamin Leary, were both missing believed drowned, but the skipper's twenty year-old son who managed to swim 200 yards to the shore and 'struggled, exhausted through frozen mud to reach a cottage', was recovering in hospital. The twenty-two man crew of the Lebanese cargo ship *Alfotis* were rescued after running aground on the Herd Sands at the mouth of the Tyne. But rescuers failed to reach the infelicitously named Sunk lightship off Harwich, where a lamplighter, R Buckland, had broken his ribs in a fall; for the first time in forty years lifeboatmen at Walton couldn't board their boat moored to the pier. A sixty year-old North Sea Pilot, Commander Albert Longmuir of Petworth in Sussex drowned off Dungeness after he fell into the sea while the Boarding the American ship Blue Jacket.

Police searching an area around an abandoned car at Belthorn near Blackburn found the body of John Yates, 48, the father-in-law of the Manchester City and England inside-forward Peter Dobing, but in the Lake District, a twenty-man rescue team working by floodlights and using ladders, ropes and a boat, rescued Mr and Mrs Curwen of Belle Isle, Bowness-on-Windermere, after 'a three hour ordeal'. They had fallen through the ice while crossing the frozen lake to their home. Ten Glasgow families also had a narrow escape when the end wall of a four-storey tenement in Barloch Street Glasgow collapsed in the gales. All of them managed to scramble to safety.

Two twenty year-old students were rescued after being buried by an avalanche between Rochdale and Edenfield but, despite a five hour search by a team of more than twenty people, the third member of their party was still missing at nightfall when the search was suspended. It would resume at daybreak but hopes of his survival were slim. Police and rescue teams also carried out a day-long search for two climbers buried under a 100-ton avalanche at Chew Valley, Saddleworth. Both men, Graham West, 29, of Stalybridge, and Michael Roberts, 27, of Dukinfield, were married and members of the Manchester Gritstone Climbing Club. The search was abandoned at midnight when no trace of them had been found. Their bodies, and that of the missing student, David Leith, 20, buried by an avalanche, were found the next day.

That day, 21 January, the last 111 civilian workers were flown out of Fylingdales by helicopter, but rescue came too late for sixty-two year-old William Bailey who was found dead on his farm at Allerton in Bradford, buried in a snowdrift. Ambulance men had to battle through waist-high snow to reach the farm and help dig out the body. Four elderly women also died in three separate incidents after the frost caused gas mains to fracture. Mrs Rhoda Thompson from Bradford, Mrs Ivy Cousins and Mrs Winifred McKeenan, widows who lived in neighbouring cottages in Seaton, Devon, and Mrs Harriet Kite of Bath were all poisoned by leaking gas. A gas-boiler explosion also injured four people at Garston near Watford. Plumber Leslie Wyatt who was working on the boiler at the time, suffered head injuries when the blast 'threw him from the kitchen along the hall to the front door'.

As the frost bit ever deeper, the Humber was frozen in places four inches thick and 'the wakes of the ships froze before they could disperse'. Southend pier was surrounded by sheet and pack ice, which stretched for more than a mile into the Thames estuary. A further attempt to take the injured man off the Sunk lightship off Harwich failed again because of continuing bad weather.

In Westmorland, the springs that normally supplied cattle herds with water had been frozen solid for eight days and reserve supplies were running out. Kendal firemen had to deliver 400 gallons to Ellenwray Farm at Old Hutton when 'forty Ayrshire cows had nothing to drink'. Human problems were eased by British Rail which, with Peak District roads 'blocked by nearly twenty-foot drifts, with cars, vans and buses abandoned', agreed to reopen the Buxton to Ashbourne Line, closed to passenger traffic for more than a year. Preceded by a snow plough, a special train pulled by two engines set out for Buxton, laden with bread, milk and other supplies.

Power-cuts were again occurring and police escorted a mobile generator borrowed from a building site in Stoke Newington to the Garrett Anderson Maternity Home in Hampstead, where 'the incubators sustaining three babies' had been affected by power-cuts that blacked out Hampstead. At Hampstead General, where another three babies were in incubators, staff asked police for a battery 'to keep the oxygen supply going'.

On 23 January the temperature dropped still further and the coldest night of the winter left thousands of diesel vehicles stranded. 200 London buses froze up on the roads, mainly in Hounslow and Barnet, and the AA reported that lorry drivers on the Barnet by-pass were 'lighting small fires under their lorries to thaw out their frozen fuel systems'. Over a hundred vehicles were abandoned on the M1, twenty more on 'a road leading in at the Birmingham end', and mist and fog made the roads even more hazardous - 'the worst driving conditions of the winter', according to the RAC, leading to a twenty-seven car pile-up in fog near Toddington. London Airport was fogbound and a Whirlwind helicopter from Manston in Kent had to airlift heating fuel to a ship frozen into the ice at the mouth of the Blackwater River in Essex. The death toll from the cold continued, with the body of a seventy-two year-old Aberavon woman found frozen in the snow, just a handful of yards from her front door.

*Only the marker posts and telegraph poles show where the snowbound moors end and the road begins.*

By 24 January there were the first hopeful signs since the cold snap began, with weather forecasters predicting a slight thaw and a modest improvement in coal and power supplies. The National Coal Board said that fourteen trains had been loaded at East Midlands pits to take coals to 'points where merchants could collect it', and 1,500 lorries took part in 'an emergency coal lift' to power stations and homes. Although the Wales Gas Board had 'stopped selling appliances as an emergency measure', restrictions on Welsh gas supplies were also eased, but elsewhere there was more bleak news. 2,200 of Oldham's 2,600 gas-lamps were out of action and most of the remaining lights were 'giving little more than a glimmer'. Freezing fog in the West Midlands dislocated electricity supplies when thick ice formed on power lines, there were more power-cuts in Kent and water rationing in Carmarthenshire because the mountain streams that normally fed the reservoirs were all frozen solid. As a result, water was cut off from 9.30pm to 5.30 am every night. Already dangerously low, supplies were further threatened when a main burst in Carmarthen, leaving half the town without water for another fourteen hours.

The football pools companies had been losing money hand over fist since the cold snap began and after three successive pools coupons had been wiped out by postponements

*Telegraph wires are bowed to breaking point by the weight of ice and a railway water tank is smothered in icicles.*

caused by the weather, they took steps to reduce their losses. For the first time ever, a "Pool's Panel" was set-up with "The Five Just Men" - retired footballers Ted Drake, Tom Finney, Tommy Lawton and George Young, and retired referee Arthur Ellis - were recruited to predict the results of games that had been postponed because of snow and ice. The panel, paid £100 each for their trouble and meeting under the chairmanship of Lord Brabazon, sat for the first time on the following Saturday, 26 January 1963.

The state of the weather suggested that the Pools Panel would be meeting regularly over the next few weeks, and there was unlikely to have been much popular sympathy for the proprietors of "England's longest ski run" at Edale in Derbyshire, who, alone among their fellow-countrymen, were complaining of a shortage of snow. They explained that strong winds during the week had blown away most of the snow on the ski-run. Elsewhere there was more than enough snow and ice for everyone. Hardraw Force, at 100 feet the highest single-drop waterfall in England, was almost completely frozen. There was 'a solid block of ice' seventy feet wide and fifty feet high at the foot of the waterfall and one of the 'giant icicles' hanging from the upper edge was sixty feet long. A cat was found frozen by its tail to the corrugated iron roof of an outbuilding in York, and frost cracked a plate-glass display inside a confectioner's shop and 'sent jars of sweets crashing down'.

'Thousands of huge ice-floes, many of them more than thirty feet across' were carried up the River Taw in Devon by the high tide. They piled up against the two bridges at Barnstaple and 'soon there was a solid mass of packed ice, many feet thick, extending right across the 500-foot wide river. On Merseyside where the daytime temperature was -11°C, two of the three docks at Garston were frozen up and in Yorkshire, the Humber Conservancy Board warned that it was 'extremely hazardous to attempt navigation of the icebound river'. All the channel marker buoys between the River Trent and Hull had been cast adrift or dragged off-station by the ice-packs that stretched from Goole to the North Sea, a distance of forty-seven miles.

Fourteen girls aged between fifteen and seventeen were taken to hospital after being 'overcome by cold while travelling in a coach from Basildon to London' and the British Medical Journal carried a warning to mothers about the dangers of hypothermia to babies. Mothers were instructed to 'feel the baby's face and extremities from time to time in cold weather'. Almost the only people feeling the heat were the fifty families evacuated from a building in Streatham after a fault in the central heating system caused 'radiators and pipes to buckle with the intense heat'. The families stood shivering in the courtyard for an hour while firemen worked to "draw out" the fire in the boiler. 'The whole building was shaking and the noise was terrifying.'

Wildlife was badly affected by the cold. Mrs Joan Sydenham, 40, was staying alone on the icebound Long Island in Poole Harbour, trying to save thousands of starving seabirds sheltering there. Foxes were raiding dustbins and 'decimating' the Chipping Norton cats and a man driving a bulldozer on the Burford road in the Cotswolds was 'picking up dead pigeons and small birds for three days or so'. For several miles along the Tewkesbury road out of Stow-on-the-Wold, skeletal rabbits and hares could be seen 'running about on the verges, ignoring the sound of the car. They are nothing bur skin and bone and I suppose come down to the road to eat the bark of the trees'. In a scene that could have come from Hitchcock's film The Birds - released that year - a woman carrying bread rolls on the top of her shopping basket was attacked by hundreds of starving pigeons as she walked across Town Hall Square in Leicester. 'As they tore at the rolls, she was knocked off her feet. She then got up and ran off.'

The appalling weather was not confined to Britain, with a savage winter right across the Northern Hemisphere. Only in parts of Scandinavia and some parts of the far North was the temperature milder than usual, with Western Norway recording a January temperature of 6°C. There were blizzards in Japan, where twenty people died in an avalanche, and in the USA, the temperature in Jackson, Mississippi, fell 12°C in a single hour, Bowling Green, Kentucky recorded a low of -29°C and Atlanta -19°C, the coldest temperature there since 1899. Further north, a fifteen-foot thick ice-floe blocked the mouth of the Niagara River, while tons of ice tumbling over the Niagara Falls formed mounds 'up to seventy feet high' in the lower river.

In Europe, 230 villages were cut off by snow in the Apennines, there was ten inches of snow in Bari, right down in the "heel" of Italy, and tourists seeking a little winter sunshine in the Sicilian town of Taormina were 'throwing snowballs at each other'. It even snowed in Tunisia. In Belgrade and Bavaria the temperature dropped below -32°C, and Swiss lakes and rivers were frozen, some for the first time since 1929. However, unlike London Airport, Zurich's Kloten Airport remained open every day throughout the long cold spell. An airport employee had devised a mixture containing alcohol and 'a secret ingredient' as an antifreeze for his car, but having tested it as an aircraft and runway de-icer, it was pressed into service at the airport. 'All the trials having been successful, the Zurich Airport had a motor tank constructed holding 1,350 gallons with twenty-four-foot hoses on either side for spraying the runways. It takes from thirty-five to forty minutes to spray the main runway and the ice melts within five to fifteen minutes.' Each application cost £100 but the airport found the process 'highly profitable, as seventy aircraft landed and paid £1600 in airport duties'.

In Denmark a cinema owner from Randers took bread rather than money as the price of admission, so that he could feed the starving wild birds, and French firemen had to rescue seagulls after their feet froze onto the ice on the River Rhone and in Germany the upper Rhine froze for the first time in thirty-four years and barges were at a standstill in Bonn. The ice blocking the river greatly reduced its flow, causing problems downstream when sea water penetrated further upstream than usual and penetrated the Maas, a Dutch tributary of the Rhine from which Rotterdam drew much of its water supply. The city's water became 'undrinkable' with a salt water content ten times the normal level, and Rotterdam grocers began selling fresh water at 3d (just over one new pence) per litre.

On 24 and 25 January the promised slight thaw made a brief appearance in Northern England and Scotland. The temperature in Bridlington jumped 12°C in three hours, but the thaw triggered avalanches on the Snake Pass between Sheffield and Manchester. After blasting operations failed to clear the snow and rock debris, it was announced that the pass would remain closed for another fourteen days. The thaw also caused 100 bursts in the water mains in Ribbleton, near Preston in Lancashire, and turned many miles of roads into 'rivers of slush'. In the South the temperature remained well below zero. At Gatwick it fell to -13°C and freezing fog caused power-cuts in London and Brighton. There were also power-cuts in Derbyshire, where five ice-cream vans used their generators to provide warmth and light for premature babies at two Derby hospitals. Luton endured an eighteen-hour power-cut after cables and condensers became frozen, and water supplies in Mid-Bedfordshire 'failed totally'. In the hope of saving water, with supplies now 'drying up all over the country', women were urged to 'forget washday' and keep their dirty clothes until the weather improved. There was a faint hope of an improvement in the power supply situation with the news that eighty shiploads of coal left the North-East for London, the South and the South-West, while 4,000 lorries were 'working through the weekend' in a 'blitzkrieg-style operation' to move coal.

Army engineers used dynamite to break up ice on the River Colne in Essex so that four ships at Hythe Quay could put to sea, and there were reports of hundreds of 'frozen suppers' washing ashore there and all along the East Coast, as conger eels 'killed by the big freeze-up' were washed ashore. Experts said that it was forty-six years - the terrible winter of 1917 - since the waters of the North Sea had last been cold enough to kill the eels. A Boulogne fishmonger's clerk would have had some sympathy for the eels. The newspapers reported that he had moved his desk to the cold storage room among the fish because it was warmer than the office. 'It's 2°C in there [the cold storage room]' he said, 'and the temperature is -6°C in my office.'

While a few eels may have been a welcome addition to some families' diet, attempts to improve the supply of other food were hampered not just by the weather but by the depredations of starving wildlife. Market gardeners were 'waging a shooting war against starving pigeons eating spring broccoli and 100 head of deer from Scrivelsby Court Park,

*Imprisoned in ice - many canals and rivers froze solid, trapping boats until the spring thaw.*

Horncastle, 'home of the Queen's Champion, Major J Dymoke, were invading neighbouring farmland and eating kale and root crops intended for sheep'.

A moderate improvement in the weather that weekend appeared to have come too late for an escapee from Dartmoor prison, Robert Crook, 34, who had absconded from a working party with another man, but was now feared dead, after police searching the edge of a plantation two miles from the prison on the Tavistock to Exeter road, found erratic 'footprints and marks in the snow, suggesting the journey of an exhausted man'. His fellow-prisoner had been recaptured unharmed.

The slow thaw strengthened that weekend but that brought fresh problems in its wake. Rail tunnels from the South-West to Kent were closed because of the danger from falling ice. The Metropolitan Water Board reported that it was 'inundated with reports of burst pipes'. Hundreds of rare books were damaged when a pipe burst in the old library of Trinity College, Oxford, and when a water main burst in Southampton Row in London, water 'swept like a wave down the street. Jets forced up paving stones and thousands of gallons poured into the site of the New Bedford Hotel'. Twenty-four hours later, firemen were still pumping out flooded basements of shops and hotels, and parts of Southampton Row remained closed for a further week. Large ice-floes - some 'four feet thick and half an acre in extent' - broke away from the frozen Langstone Harbour near Portland and tore away the anchors and cables of more than a dozen yachts and cabin cruisers, setting them adrift.

Insurance companies added to the general air of gloom by announcing that they were facing tens of thousands of claims totalling an estimated £15 million, 'the most expensive winter ever'. Farmers in East Anglia had their own financial woes to worry about, as the British Sugar Corporation reported that sugar beet still frozen in the ground was now 'likely to be unsuitable for sugar manufacture'. Farmers could not even use the spoiled crop as fodder for their animals as 'it may be dangerous... we recommend that expert advice should be obtained before doing so.'

The month of January had been one of the most miserable in British history - the coldest month of the entire twentieth century. Mean minimum temperatures were well below freezing almost everywhere in England, Wales and Scotland, and even the mean maximum temperatures were at or below freezing in many areas. Between 23 December 1962 and 25 January, 1963, the weather station at Moor House in Westmorland - a relatively low-lying station - recorded thirty-four consecutive days with a maximum temperature of zero°C or below. At higher altitudes, the temperature were markedly colder, with Braemar recording a low of -22.2°C on 18 January. There were air frosts throughout southern England and South Wales on 25 of the 31 nights during the month, and even at normally mild Kew, the temperature never rose above 2.7°C at any time in the whole month. The CET (Central England Temperature) has, since 1659, recorded the monthly average temperatures for an area roughly bounded by Preston, Bristol and London, and in January 1963, the CET of -2.1°C, was 4-5°C below normal, and even lower than the -1.9°C recorded in February 1947.

The weather had helped to drive up unemployment, which 'reached new peaks' in January, while industrial production had fallen nine per cent below the November level. The one modest consolation amongst this catalogue of woe was that between the blizzards, the high pressure and clear skies that characterised the savage frosts made for sunny days. Weather stations in the West, from the Hebrides, through Prestwick, Manchester and Aberporth to Chivenor, recorded twice the normal monthly average of sunshine hours.

The change in month brought no let-up in the weather, with the return of fog and snow marked by falls of two inches in Essex. The emergency 'coal-lift' to the South continued with around one hundred 'trimmers and teemers' in Sunderland working in blinding snowstorms to load 10,000 tons of of coal into ships bound for the Thames. In the East Midlands another 4,000 lorries took part in a second 'emergency lift' of 50,000 tons. Most went to power stations and to homes in the London and Bristol areas, but a Bath coal merchant 'recruited 400 lorries' from as far away as Weston-super-Mare and Bristol. 'In 100 years, there has never been so much coal distributed in one day in Bath.'

On 5 February, rubbing the noses of his fellow Britons in their winter misery, Commodore FG Wallis, the new master of the Queen Elizabeth, remarked that the liner's voyage to New York had been 'practically summer cruising', the noon temperature on arrival in New York being 14°C. As he made his remarks, gales and yet another blizzard were sweeping his home country. More than 130,000 miles of roads were affected. There were gusts of seventy-five miles an hour at Ronaldsway, Isle of Man, and sixty-nine miles an hour in Blackpool. Two elderly sisters, Misses Ellen and May Saunders, were found dead of hypothermia in their caravan at West Kingsdom, Kent. The West Country was again almost cut off from the rest of the country, with not a single road open into Cornwall. After 'some of the worst blizzards in the memory of many people in South Wales' eighty men assisted by six snowploughs finally succeeded in digging their way through to an ambulance stuck in the drifts between

Aberaeron and Aberystwyth. It was carrying farmer David Williams, 26, of Lampeter, who had been seriously injured in a tractor accident. Fire Brigade officials advised the Thames Valley Water Board to 'discontinue the method of sending electrical charges through pipes to thaw them'. The warning came after the method caused fires in two houses in Reading. The Water Board refused to discontinue the practice but announced that 'water engineers will be accompanied in future by an electrician and a gas fitter'.

The next day, the editor of The Times had had enough of winter and its attendant problems and complained in an irritable editorial that 'some deaths, much misery and general inconvenience are the prize the nation has paid since Christmas for being unready' for the harsh weather. He went on to lambast in turn the water, gas and electricity supply boards, the railways, national and local government and 'last but not least in the catalogue of ineptitude, the strongly entrenched British tradition of jerry-building in plumbing. Cisterns and pipes are installed in our island homes as a sacrifice to the ice gods. Their placing is such that often blockage or burst is a near certainty in a bad winter.'

Whether through the intervention of The Times or the ice gods, a thaw set in the next day, bringing the inevitable floods in its wake. Boscastle in Cornwall was flooded when thousands of gallons of meltwater swept debris down the valley and blocked the River Jordan. The thaw continued the next day, 7 February, with the River Exe bursting its banks between Stoke Canon and Exeter. 15,000 sandbags, blankets and radio equipment were being stockpiled in Devon. Troops were on standby and Army engineers used dynamite to

*An old steam tug, Mayflower, leading the tanker barge Keeldale H through the ice on the Gloucester and Sharpness Canal in January 1963. The ice was so bad that the tug had to be brought out of retirement to help with the job of keeping the canal open.*

*The tug Primrose hauling the first of a line of ships past the Quedgeley oil terminal on the Canal. Convoys of ships were formed to try and keep the canal open.*

*A car swept away by floods*

break up 'huge ice-floes' on the Rivers Axe and Exe, and to blow up an ice-jam that was backing up flood waters threatening to sweep away the main Exeter to Lyme Regis road. In Wales, a fifty year-old man from Swansea was drowned in the river Amman in Carmarthenshire when he fell while trying to break up an ice-floe on the river.

While the thaw gathered pace in the South-West, Scotland was experiencing its worst snowfalls of the winter. A Wigtown snowplough was 'struggling to take supplies' to six adults and eleven children trapped in a school and adjoining house at Meoul near Portpatrick. An RAF helicopter from RAF Leuchars, 150 miles north-east of the school, had been forced to turn back and another attempt to take food to railwaymen stranded at Riccarton Junction, Roxburghshire also had to be aborted. Of the cross-border routes, only the A1 and the Dumfries to Carlisle roads were still open. Nineteen passengers on the 1.25pm train from Mallaig, Inverness-shire to Glasgow were stuck all night in snowdrifts between Bridge of Orchy, Argyllshire and Tyndrum, Perthshire. They eventually reached Glasgow after nightfall the following day. Passengers on the 4.25pm train were also heavily delayed; among them, appropriately enough, was Colonel Donald Cameron, chairman of the Scottish Railway Board.

In England a bus carrying thirty-five schoolchildren plunged into a dyke and overturned after it collided with a car on an icy road near Sleaford in Lincolnshire. 'The children had to scramble on top of the bus after the crash and jump eight feet into frozen snow to safety'.

Miraculously only three children were injured and only one of them hospitalised with a broken arm. A helicopter crashed about ten miles from Darlington shortly after taking off during a snowstorm from the RAF station at Middleton St George. The pilot 'decided the weather was too severe and tried to land in a field, but the helicopter lurched over and cartwheeled for 100 yards on its rotor blades.' Both the pilot and his passenger scrambled out unhurt. An Army helicopter, one of seven carrying out missions to isolated farms in Northern Ireland, also crashed near Mossley, County Antrim, but once more the pilot escaped unharmed.

In response to emergency appeals from all parts of the country, the Red Cross sent out '500 blankets and fifty cartons of soup to Glamorgan for sick old people; coconut mats to Penzance; scrubbing brushes, floor cloths and disinfectant to help in clearing up flooded homes in Montgomeryshire; blankets to Halifax, Yorks; and babies' napkins, blankets, pillowcases, floor cloths, mops and soap to Northern Ireland.' The purpose of the coconut mats was not explained.

The next day the immediate threat of serious flooding in the South-West had eased a little, though ice was still jamming bridges and 'the long-term situation was still critical'. Further north, the Scottish borders were still snowbound with a five-mile queue of vehicles on the A76 from Cumnock to Dumfries 'trying to grope their way south'. Bad weather again prevented an air-drop of supplies to the adults and children trapped in the Wigtownshire school, but reporters from the BBC and three rival newspapers 'tramped across the snow-covered fields, taking bread and tinned meat' to them.

On 9 February, a team of quarrymen employed by Derbyshire County Council mad the 700-foot ascent of the mountain overlooking the Snake Pass in the Peak District and used gelignite to blast away a twenty-five foot overhang of snow threatening the road. It was so cold that the eyelashes of some of the men froze shut. After the blast, 'giant bulldozers and a snow-blower then went into action to sweep away the avalanche and the tremendous task' of reopening the pass was finally completed. However, further snow fell continually all the following day, blocking the pass to traffic once more. In Essex the AA reported that the temperature was 'diving' turning roads into 'glittering sheets of ice'. Cars were 'skating' along main roads and traffic on minor roads was virtually halted.

A few days respite from the worst of the weather followed, but on 15 February winter returned with a vengeance. There were heavy snowfalls from Hampshire, Wiltshire and Dorset in the South, through Derbyshire, Lincolnshire and Yorkshire to the Moray Firth and beyond. Railway lines from Glasgow to Fort William and Edinburgh to Carlisle were blocked, and conditions on the A9 Perth to Inverness road were described as 'the worst of the winter', with cars and lorries completely buried in fifteen-foot drifts near Kingussie. In England, roads in the Peak District and Yorkshire Dales were impassable and in Wales, roads between Brecon and Swansea and Merthyr were blocked. In the North-East, sheep rescued from snowdrifts were found to be 'suffering from snow-blindness', according to Mr EE Milner of the National Farming Union. 'They have been taken under cover and are recovering.' Further south, floods were now affecting new areas with waters rising rapidly in the Stour Valley and Blackmore Vale areas of North Dorset and East Somerset, where hundreds of acres were already under water. Children had to be rescued from several schools by farm tractors. As if all this was not enough, MPs found time to berate the Air Ministry for its decision to begin giving temperatures in Celsius instead of Fahrenheit; there

*Floodwaters roaring through the appropriately-named Canal Road in Bradford.*

were complaints about public 'disgust, unhappiness and inability to follow temperatures given by the Air Ministry in Centigrade [Celsius]'.

Whatever system was used, temperatures remained well below freezing in the North, where two men's attempts to provide relief for swans on the frozen River Wharfe in Otley, Yorkshire, only earned the disapproval of the RSPCA. 'Armed with pickaxes and hatchets', they broke holes in the ice-covered river but were told by RSPCA Inspector Albert Drew, 'You are doing more harm than good.' The deaths of two of the sixteen swans on the river had provoked 'irate animal lovers to contact the RSPCA to ask "What are you doing about it?" They received an equally straightforward reply: "Nothing".' Inspector Drew claimed that 'in all probability the two swans had died because they were attacked by rats' and said that he had paid several visits to Otley during the past few weeks, 'and I don't think there is anything wrong with the swans. They are quite comfortable on the ice and well fed. If people go breaking the ice haphazardly, it will be more likely to freeze over completely because of the small ice-floes, and swans resting in the water could be trapped.'

A far more serious threat to wild birds was highlighted by news of 'an alarming decline' in the numbers of wildbirds in general and birds of prey in particular - the "Silent Spring" about which Rachel Carson had written in her seminal book published in 1962. Concern was fuelled by reports of 'thousands of birds that fell out of the sky in agony' near Caistor in Lincolnshire. An official at Caistor Hospital 'where hundreds of birds fell' reported that, 'These birds were screeching in pain and I killed several myself'. RSPCA officials took some of the dead birds away for a post-mortem but no obvious cause was found. Rachel Carson and many others felt that the most plausible cause for such incidents and the decline of wild birds in general was the number of toxic chemicals, including DDT, being used on the land. However, a Ministry of Agriculture spokesman was at pains to claim that there was 'no cause for alarm, provided chemicals are used discriminately. We have a voluntary agreement with the farmers not to use certain chemicals at certain times of the year.' DDT did not seem to be affecting a million-strong flock of starlings roosting in a nine-acre wood in West Cumberland. 'Gunshots have little effect on the birds and they ignored blasting operations on some nearby rocks. When they are settled in, they make so much noise that it appears that the explosions affect only birds on the fringes of the flock.' At dawn, 'the birds take off together like a mushroom cloud from an atomic bomb and disperse before returning to the desolate wood at nightfall'.

With disruption on the railways now easing, there was an ominous warning that the cuts to the rail network proposed by "the butcher of the railways" Dr Beeching, would be 'more severe than the most pessimistic estimate so far, because lines will be closed by the hundred'. It was hardly the ideal moment for Dr Beeching to show off a new range of uniforms for British Railways staff, albeit long overdue, the first redesign in fifty years. Sydney Greene, General Secretary of the National Union of Railwaymen, gave the redesign his grudging approval, saying 'Railwaymen are the worst-dressed public servants, so anything is an improvement.'

The worst of the winter had now passed and there were no more heavy snowfalls, though temperatures remained very low with harsh night frosts throughout the rest of the month. The snow still lying in the North proved a boon to police who used 'Sherlock Holmes-type detective work' to arrest a thirty-nine year old labourer who had attacked a teenage girl. He left a trail of 'diamond-patterned heelprints in the snow' that police followed to his front door. Newspaper sub-editors desperate for new slants on the winter weather must have been saying prayers of thanks for the story that broke on 26 February when Maxwell Fletcher, 17, a Dublin cadet officer who had been at sea only two months proved the hero of a fire aboard the British tanker Abadesa in the River Scheldt. Although 'covered in ice', for ninety minutes he remained at his post on the deck of the blazing ship - prompting endless tabloid variants on 'the boy stood on the burning deck' theme - directing the jet of a hose into the fire threatening the ship. Chief Officer T Woolcott of Bournemouth said, 'It was about ninety minutes before he was relieved and by then he was solidly covered in ice, but not complaining. His chest seemed to be a solid block of ice.'

If not quite as cold as January, February had been exceptionally bitter. Snow fell somewhere in the country on twenty of the first twenty-three days of the month, and during the whole of February, County Durham experienced an air frost every single night. Kent and Essex went eleven consecutive days with unbroken frost and, taken together, the three months - December, January and February - constituted 'the coldest winter in the London area since before 1841'.

In the last days of February and the first days of March 1963, sunny weather raised afternoon temperatures to four or five degrees, but the clear skies saw temperatures plummeting again at night, and it was not until 4 March that mild south-westerlies brought rain and rising temperatures. The warmth led to 'Spring fever' on the roads with the RAC estimating that 15,000 vehicles an hour were leaving London for the coast.

On 6 March, the daytime temperature in London reached 17°C - the highest since 25 October of the previous year - and the night that followed was the first frost-free one right across the country. With the thaw came more flooding, but on nothing like the scale of the 1947 floods, and the Thames Conservancy deliberately lowered the water level in the river to break up the remaining ice that, weakened by the thaw, was a serious hazard to children venturing on to it. As if in emphasis of that, twin boys wearing cowboy hats disappeared after falling into the swollen River Wear at Sunderland, a few yards from their home. The body of one of them was retrieved from the river later that day. And an eight year-old boy was missing believed drowned after being swept away by flood waters at Newton Aycliffe, County Durham. Malcolm Walker 'put his foot in the water to see how deep it was and suddenly slipped. Then the current swept him away.' The boy and his brother had been warned by their primary school headmaster earlier that day not to walk home by the beck because of the flooding. Police called off the search for the missing boy at nightfall.

Ice-floes 'as big as dining tables' swept twenty miles down the River Tyne, 'causing chaos among small craft'. On the River Ouse in York people 'gathered to watch the ice crash into the moored barges' faced another danger when 'the concrete bordering of the bank, loosened by the frosts', began cracking. And the A66 Penrith-Scotch Corner road was closed at Brough after it collapsed; 'the road sagged and holes began appearing'.

As the winter came to an end, government and people began to count the cost in lost income and extraordinary expenditure. Household fuel bills had skyrocketed, many factories had been on short-time working or temporary closure because of power-cuts. Many farms in remote upland regions had been isolated for over two months, tens of thousands of sheep had been lost and crops worth millions of pounds lost or severely damaged. There was a sharp increase in unemployment in the building trades, and a steep rise in insurance

*Canoeing enthusiast Clive Baker took time off from his job as Halifax Town coach to paddle his own canoe - on the pitch. Floods covered part of the ground to a depth of two feet and Clive took the opportunity of getting in some useful practice.*

premiums. Local authorities also faced a huge bill for road repairs after frost-heave, as well as the short-term costs of keeping the roads clear - five million cubic yards of snow were moved from Gloucestershire roads alone, and the North-West counties estimated their total bill because of the severe weather at in excess of £1 million, and Lancashire County Council alone spent £400,000 keeping the roads clear.

The winter of 1962-63 was over but, if not the longest, or the snowiest - snowfalls were heavier in 1947, and the depths of lying snow even greater - but 'from the point of view of large areas of the country being under snow, this has been the worst winter since 1814'. And 1962-63 was to remain the coldest winter of the twentieth century and the second coldest ever recorded - only 1739-40 has ever been worse.

ROAD
CLOSED

**PART IV**

# 1978-79 - The Winter of Discontent

If 1947 and 1962-63 were snowier and colder respectively, probably no winter in history inflicted more misery on the hapless British population than 'The Winter of Discontent' of 1978-79. The dominant themes of the winter were established early and every subsequent deterioration in the weather seemed to be matched by a worsening chill in the climate of industrial relations.

There had been spells of snow in November and early December, and rainfall in December was three times the normal average in the Midlands, the East and North-East, and London had its 'wettest December since records began', but the real force of winter was not felt until Christmas was approaching. There were sharp frosts in the week before Christmas and on the last day before the holiday, Friday 22 December, the newspapers headlined the threat of a strike by tanker drivers planned for 3 January. 10,000 troops were reported to be on standby to take over delivery of fuel and the RAC issued a condemnation of the panic buying of petrol that was already leading to shortages in some areas. BBC television and radio were blacked out by strikes, electricity and gas price rises of eight per cent were announced and, just to add to the festive cheer, supplies of the traditional accompaniment to the Christmas turkey - sprouts - were in very short supply. The heavy frosts 'which turn sprouts black' had stopped picking on Tuesday and Wednesday of that week and much of the cauliflower and 'open field celery' crops had also suffered frost damage.

It was not a white Christmas in England and Wales, but there were heavy falls of snow and rain in the North of Scotland on 23 December and the snow spread slowly southwards, though at first the only roads badly affected were in the Dundee and Perth areas. However,

*The two most familiar words in the English language that winter. It was often impossible to tell where a road ended and the fields began.*

the Cairngorms already had a good covering from the earlier snows and the fresh falls triggered avalanches that killed two climbers on a Christmas expedition. On Boxing Day, the bodies of Ian Kershaw, 35, of Oldham, and Colin Shaw, 28, of Leeds, were found in a corrie after a day-long search.

The papers the next day reported a speech by the Shadow Chancellor, Geoffrey Howe, in which he described 1978 as 'the year of the bloody-minded', citing the strikes, go-slows and work-to-rules that at various times had blacked out the BBC, ITN and YTV, stopped trains running on Wednesdays, dried up the petrol pumps, brought the criminal justice system to a halt with prison officers and lawyers in disputes, stopped the presses of The Times, seen firemen and bakers on strike, social workers walk out, Cheshire Homes picketed and patients turned away from hospitals. Those who thought things could not possibly be worse in 1979 were in for an unpleasant surprise.

Over the next few days the snow continued its southward journey and there were falls of six to seven inches in southern Scotland and North-East England. Ahead of the snow the temperature dropped still further and there were savage frosts across England and Wales. On 29 December the winter really began to bite in England. Snow blocked roads throughout Scotland and the North of England and combined with torrential rain to cause serious flooding in Yorkshire. Reservoirs overflowed, Dales rivers burst their banks and the River Ouse rose sixteen feet above normal levels. 'Newly enlisted soldiers' evacuated 500 families in York and laid 3,000 sandbags as flood waters five feet deep covered a large part of the city. Exceptionally high tides in the Humber later in the week threatened more flooding. The Mersey also overflowed in Warrington and the Trent was ten feet above normal and still rising. Meanwhile the AA reported that 600 miles of Pennine roads north of the Peak District were blocked by snow, with only the M62 still open. A County Durham family trying to drive to Cumbria were trapped for ten hours in a snowdrift. Most cross-border roads were also closed, though constant snowploughing kept the A1 open. Blizzards cut off hundreds of small communities in northern Scotland. At Cumbernauld near Glasgow, an elderly man died from exposure after collapsing in a snowstorm. He was found 'on a footpath in a deserted recreation area'.

The rain in the south turned to sleet and then snow and there were further blizzards on 30 December and New Year's Eve, causing the cancellation of many planned celebrations, though the traditional New Year's Day cricket match between the village of Burley in the New Forest and the appropriately named 'Jack Frost XI' went ahead on a pitch cleared of several inches of snow. 'Bottles of whisky, Afghan boots and hot water bottles' helped to keep the cricketers warm. Meanwhile, the AA was reporting that 'The picture is very grim. The snow is covering the whole of Great Britain for the first time since 1963.' In Yorkshire, helicopters rescued dozens of people including forty coach passengers stuck in drifts near Middleton-in-the-Wolds. The Glenlomond mental hospital in Kinross-shire was snowbound and local farmers and volunteers struggled through the drifts with essential supplies.

Roads were blocked from the North of Scotland to Kent, where 500 tourists were unable to reach Dover to catch their ferries because every road was snowbound. At one stage farmer Giles Hadfield 'had 180 marooned men, women and children taking shelter in our house. Their cars and coaches were trapped in drifts eight to ten feet high. During the night we made hundreds of cups of tea; our New Year drinks had gone by 6pm.' 300 passengers were stranded on two trains at Bonneybridge, Stirlingshire, for nine hours, and a hundred

*It's an ill wind... the weather brought a bonanza for breakdown trucks and for plumbers, who were kept busy repairing burst pipes well into Spring.*

passengers on the sleeper from London Paddington to Penzance were trapped by 'a combination of engine failures, frozen points and the weather'.

New Year temperatures had fallen to -7°C at Heathrow, -9°C at Gatwick, -11°C at Benson, Oxfordshire, -16°C at Crawford John in Lanarkshire and -17°C at Linton-on-Ouse in Yorkshire. Flights to and from Heathrow were delayed by up to six hours and British Airways cancelled thirteen long-haul services.' Snow now blanketed much of the Northern Hemisphere, with the Skane district of southern Sweden declared a 'catastrophe area'. In West Germany, after eighty hours of continuous snow, even army tanks trying to deliver generators to isolated villages became stuck in snowdrifts. Many parts of the United States experienced blizzard conditions and power-cuts affected more than 100,000 people in Dallas, Texas. Undeterred by the appalling weather, hundreds of Muscovites, 'armed with bottles of champagne and vodka', braved temperatures of -35°C to celebrate New Year in Red Square.

*The motorist's nightmare - drifting snow and uncleared roads.*

In Britain, fifteen London hospitals closed their casualty wards and emergency admissions because of 'staff shortage caused by the weather and sickness'; and more than seventy patients, some seriously ill, were evacuated from two Birmingham hospitals after burst pipes flooded wards. Medical staff in County Durham were having to walk 'up to eight miles' through the snow to reach work because bus services had been halted by the icy conditions, and in Northumberland, farmers were using tractors to ferry nurses through thick snow. Dorset Ambulance Service announced that 'with crews on the point of exhaustion after eighteen hours continuous duty', the service would be restricted to emergency calls only. The Chief Ambulance Officer, John Wilby, said that the weather and 'an acute shortage of petrol' meant that the transport of patients would henceforth have to be 'reviewed daily'. Glasgow ambulances were also only responding to emergency calls.

The RAC called for 'a full investigation into gritting operations throughout the country', after reports that many main roads were not gritted or salted despite forecasts of snow. 'We realise that some local authorities were faced with industrial [action] problems and were unable to get the gritting machines and snowploughs out, but we are anxious to ensure that adequate resources are available to cope with these situations.' Impressions of government competence were not helped by the news that Environment Minister Denis Howell had called Prime Minister James Callaghan at home to keep him informed of efforts to clear the roads, but 'no one answered the telephone at the Department of the Environment in London.'

By the next day, Mr Howell was fighting back, claiming that in a temperate climate it was not possible 'to have a whole army of men standing by to clear the roads' and saying that local authorities had 'coped quite well' with the snow. A 'Central Operations Unit' at his own ministry had started functioning 'within half a day' of the first reports of heavy snow. Unimpressed, the RAC said that it would be carrying out its own inquiry and claimed that many roads from Manchester and Essex had not been gritted at all, and that gritters in Lothian, Scotland, were not even due to return to work until Thursday 4 January.

The death toll from the weather was climbing rapidly. The body of a seventy-seven year-old man was retrieved from a river in Dulverton, Devon, and Mrs Phyllis Baillie, a widow from Culross, near Dunfermline, was found dead after a fire at her home. Firemen on their way to the blaze were trapped in snowdrifts and though some of them struggled through the snow on foot, Mrs Baillie was already dead when they reached her home. Eighty-six year-old Robert Gunn and his eighty-four year-old wife Jean were found dead in the unheated bedroom of their Glasgow flat. There were further casualties of the snow and ice. At Stanford-le-Hope in Essex, a man was found dead, covered with snow in a car park. Sixty-two year-old Stanley Barriball, 'an eccentric bachelor who always refused to wear a jacket', was found dead 'in his shirtsleeves in a lonely lane' near his home in Lawhitton, Cornwall. A twenty-five year old policeman, William Pepperell, married for only a year, was killed when his car skidded on an icy road and ploughed into a tree at Dudley in the West Midlands.

In London, three men, David Ennisen, Tony Gray and Charalambos Kourdolis, 'all aged twenty from the Kentish Town area', died when the ice gave way as they were sliding on the frozen "Gentlemen's Bathing Pond" at Hampstead. Two other men, including Mr Gray's brother Laurie, scrambled to safety out of the icy water with the help of rescuers but efforts to rescue the others failed. Martin Cooper, 21, a joiner drinking in a nearby pub, the Freemasons Arms, ran to the pond and pulled himself across the unbroken ice before plunging up to his waist in the frigid water. 'It was unbearably cold,' he said. 'I grabbed a man and manhandled him towards a tree stump where he clung and heaved himself out'. Firemen and police were joined by park-keepers who took boats out to search for the missing men.

At sea, six people, including a girl of twelve, had to jump into the Humber lifeboat from the Dutch ship Diane V after the cargo shifted when the ship was struck by a huge wave. Four fishermen from Benfleet in Essex were missing, believed drowned, despite an intensive search by police, coastguards and an air-sea rescue helicopter, and two crewmen of the trawler Ben Asdale were swept to their deaths after it ran aground in a blizzard near Falmouth. The engineer of a Russian factory ship who was helping to repair the trawler's steering gear was also drowned. A thousand people were evacuated from their sea-front homes in Jaywick, Essex, after a freak tide breached the sea wall. At Walton-in-the Naze a 100-foot section of the pier collapsed, and at nearby Clacton, a sixty-foot wave caused a quarter of a million pounds worth of damage to the lifeboat station at the pier. "Suzy Wong", a killer whale kept in the pier's aquarium had to be rescued from 'the fury of the sea' and was taken by road to Windsor Safari Park'. A Clacton police spokesman warned of an additional hazard to sea-front sightseers - flying ice. 'Spray from the waves is freezing in the air and razor-sharp slivers of ice are being hurled onto the promenade.'

As Britain began crawling back to work there was the prospect of a fresh round of industrial disputes. Leaders of four public service unions were meeting to discuss co-ordinated action in support of their claim for a £60 a week minimum wage for nearly 1.5 million manual workers and the National Coal Board was considering its reply to the National Union of Mineworkers' forty per cent pay claim. Fuel tanker drivers began their strike, as planned, on 3 January, and lorry drivers started unofficial action, bringing immediate warnings of effects on food supplies. A Road Haulage Association spokesman said that 'within a short time there could be very little food in any shops' - a self-fulfilling prophecy that virtually ensured a wave of panic buying. Coal and milk deliveries were also threatened and it was claimed that the stoppage by lorry drivers delivering newsprint to national and provincial newspapers could 'bring the whole of the newspaper industry to a halt within days'. To complete a miserable New Year it was also announced that people with burst pipes faced waits of several weeks before plumbers could repair them because an industrial dispute at the UBM Mercian builders' merchants had caused 'an acute shortage of copper piping'. Not everyone found the industrial chaos inconvenient; 50,000 children returned to school after the Christmas holidays on 3 January were sent home again the following day because of shortages of heating fuel caused by the strike by tanker drivers.

London commuters returning after the holiday found one third of peak-hour train services had been cancelled. Southern Region's catalogue of woes included 'frozen points, damaged engine cooling systems, staff shortages and rolling stock being in the wrong place.' A French doctor made light of such minor impediments. Finding that he could not drive the twenty-four miles from his Calais home to his work in Boulogne because of snow and abandoned vehicles blocking the roads, he caught the cross-Channel ferry to Dover, and then travelled back across the Channel on another ferry to Boulogne.

He was fortunate not to have been taking the same route to work later that week for on 4 January 'the worst sea storms and onslaught in living memory', hit the Channel, battering the South Devon villages of Torcross and Beesands. Mountainous seas, driven by Force Ten gales, 'smashed through windows and walls, reducing some buildings to rubble and floating planks of wood'. Houses were left with their roofs 'flapping like paper in the wind'. Fifty foot waves 'picked up massive boulders and hurled them at the helpless seaside communities'. 200 villagers were evacuated from Beesands as 'weeping women recalled that the nearby "sister" village of [South] Hallsands was completely engulfed and destroyed by the sea in 1917.' A police spokesman added 'We have all aged noticeably down here in the past few hours. Conditions there are appalling, the road between the two seaside villages has collapsed and we have immense difficulties getting essential vehicles into the area.' The route was being made 'additionally hazardous by lorries laden with boulders' to bolster the sea defences. Four loads were taken into Beesands the previous day 'and it is these the sea is

*After weeks of harsh weather, the enthusiasm of amateur snow-clearers began to fade.*

hurling back'. Boulders had been piled along the sea-front only yards away from the houses, but 'they stood little chance against waves which towered even over the highest buildings'.

A nineteen year-old Greek seaman, Demetrius Petriphs, was the only survivor of the fifteen-man crew of the 3,000 ton cargo ship Cantonard, which sank in mid-Channel off Torbay at the height of the storm. 'The seas had ripped away the weather bulkhead of the bows. I was with the first and second engineers and my mate when we heard the rescue helicopter. We all cheered when the winch came down. I grabbed for it and the friction burnt my hand. I dropped back into the water and suddenly all three of my mates disappeared. I never saw them again.' Mimi Hutchings was lucky to escape with her life after getting away from snowbound Dartmouth by 'hitching a lift' on a French yacht which was then caught in the same storm forty miles off The Lizard in Cornwall. The gales shredded the sails and wrecked the steering gear and the French owner was then washed overboard. With the yacht close to sinking, it was spotted by a merchant ship which radioed for help and Ms Hutchings and the Frenchman's distraught wife were lifted off by an RAF Sea King helicopter moments before the yacht went down.

Elsewhere a furious row over uncleared ice and snow broke out between the British Airports Authority and the seventy-five airlines using Heathrow, led by British Airways, which cancelled all its 350 flights, affecting 18,000 passengers. At a 'strong meeting' the previous evening, BA called the aircraft stands and other areas 'too dangerous to use' and complained of damage to its aircraft and injuries to its staff because of the ice and snow. Some of its aircraft had been trapped on their stands at the airport since New Year's Day. Pan American Airlines' London Director, Martin Sugrue, added that they were 'considering legal action against BAA for negligence and dereliction of duty... There have been near-blizzard conditions in Europe but nowhere has Pan American been experiencing the problems it has at Heathrow.' BAA denied negligence and claimed that in the authority's judgement 130 of the airport's 151 aircraft stands were safe to use. Priority had been given to clearing runways and taxi-ways but by the time this had been done, the snow on the parking stands and around terminal buildings had become 'compacted by the movement of aircraft and vehicles and frozen solid by continuous low temperatures'. As a result, none of the authority's equipment, 'including sixteen snow-clearing vehicles, seven gritting machines and two de-icing lorries' could be used. However, shortly before dawn, BAA brought in 100 labourers and five bulldozers to clear the 'danger areas'.

The situation would have been worse had fresh blizzards expected to hit southern counties not been 'deflected south by cold air over Scandinavia' to ravage the Channel Islands and Northern France instead. In East Germany the frost and snow left 300 goods trains snowbound, a situation the Minister for Coal and Energy, Herr Ziergiebel, described as 'a catastrophe'. Street lights in East Berlin were dimmed or extinguished to save power and troops from the twenty-two Russian divisions based in East Germany were called out to help dig people out of stranded vehicles and isolated villages. Soviet television broadcast appeals to save fuel, gas and electricity after temperatures in Moscow fell to -39°C, the lowest point in 100 years. Some areas around Moscow were reported to have been without power and water for two days.

As in previous harsh winters, newspapers were also quick with reports of unseasonably warm weather elsewhere. An Antarctic 'heat wave - relatively speaking' had seen temperatures rise to 10°C at the South Pole, Scott and McMurdo Sound bases, the highest

*Construction companies paralysed by the big freeze, hired their earthmoving equipment out as makeshift snowploughs.*

temperatures ever recorded there. Back in frigid Britain, there were further reports of fatalities caused by the weather. An eighty-four year-old widow, Mrs Millicent Smith, was found 'dead on a carpet of ice' in the kitchen of her home in Westcliff-on-Sea in Essex. It was believed that she had collapsed from the cold after turning on her kitchen tap, and the water spilling from the sink had frozen over the kitchen floor. The body of a forty-two year-old man, Raymond Hayes, was found in the snow not far from the Coniston Hotel in Sittingbourne, Kent, where he worked, and several fatalities were recorded in car crashes caused by the icy roads.

The next day, a change of direction in the wind brought calming seas and sunshine, and allowed inhabitants of storm-battered Torcross and Beesands to assess the extent of the damage to their villages and homes. As bulldozers shovelled hundreds of tons of shingle out of the streets, 'convoys of lorries of every shape and size' brought 30,000 tons of rock from Plymouth quarries to repair and strengthen the ravaged sea-defences. Several houses had disappeared or been reduced to heaps of rubble. 'A boulder that must have weighed at least a ton was thrown out of the sea', reducing The Start Bay Cafe 'to a heap of iron and stone'.

The owner, Paul Stubbs, who evacuated his wife, baby and seventy-four year old mother from the next-door pub that they also ran, returned to find the contents of that strewn along the sea-front. 'There was a billiard table across the road and two freezers in the street'. All they managed to salvage was 'a few pans and kettles'. One elderly couple at Beesands had refused to leave and sat out the storm unharmed in their tiny sea-front cottage. 'I have lived here all my life,' said seventy-eight year-old John Steel. 'I knew it would be all right in the end.'

Things were getting back to normal too at Heathrow, after 'a large workforce spent a second day and night clearing snow and ice from aircraft parking stands.' But the Saturday sports programme in Britain was 'almost wiped out', with only five of thirty-two FA Cup games 'having any hope of being played', the only rugby union game played was the Scottish trial at Murrayfield, and the only race meeting in the British Isles was at Tramore in County Waterford in Eire.

On Monday 8 January a long-awaited thaw had begun, but it only brought with it problems from burst pipes and flooding. Large parts of Sussex were flooded as dykes around Rye in the Romney Marshes overflowed and the River Ouse burst its banks between Eastbourne and Seaford. If the weather was turning a little more benevolent, the industrial situation was deteriorating rapidly. Pickets from the striking lorry and tanker drivers had succeeded in closing all the country's ports to lorry traffic - prompting the Leader of the Opposition, the little known Margaret Thatcher, to threaten laws to curb union powers - and car drivers queuing at the few petrol stations where there was any fuel to be had, were reported to be paying £5 for three gallons - twice the normal price. Greater Manchester's buses were kept off the roads all weekend to save fuel, 30,000 children in Northamptonshire were told to stay away from school and it was a similar picture at schools throughout the country. The National Consumers' Association warned housewives panic-buying food that they were 'cutting their own throats; panic-buying causes shortages and shortages push up prices.' The news that 'many grocers and supermarkets will be forced to close next week because of dwindling food supplies' was scarcely calculated to improve the situation. A National Farmers Union spokesman was meanwhile warning of 'famine' on British farms and widespread livestock deaths if deliveries of animal feed were not resumed by lorry drivers.

The next day tanker drivers voted to accept a pay offer of thirteen to fifteen per cent, but 180,000 road haulage workers moved closer to all-out official strike action after the employers refused to raise their 13.2 per cent pay offer, and the train drivers added to the chaos by announcing that they too would be undertaking industrial action from 16 January. Schools already hit by shortages of heating oil were now being closed by burst pipes and flooded classrooms. 900 schools were shut, with Sussex, Surrey, Buckinghamshire, Humberside, Lincolnshire and Yorkshire the worst affected counties.

The slow thaw continued in the South though north-westerly gales brought torrential rain and hail, but in the Midlands, North and Scotland it fell as snow with the fierce winds

*Snow-blowers imported from Scandinavia were a new weapon in the road-clearers' armoury but even they were powerless against the biggest drifts, which could only be moved by JCBs or dynamite.*

whipping up fresh drifts. Eighty-three Glasgow buses were trapped in drifts at one point. The Selsey lifeboat was called out to rescue nineteen crew and a woman passenger from a Panamanian ship which lost a propeller in the storm.

By the next day there were reports that drivers in the Midlands were 'abandoning their cars, not because of snow, but because there was no petrol to be had at any price. Even though most tanker drivers had now returned to work, Texaco drivers were still out on strike and were successfully picketing the other fuel depots in the region. Those turning to the train instead were warned that talks between British Rail and the train drivers' unions had broken down, guaranteeing that the strike set for 16 January would go ahead. Taking to the air provided no refuge either; Glasgow airport was closed because the lorry drivers' strike had prevented deliveries of de-icing fluid and it had now run out.

Some even more precious fluids were in short supply in Scotland on 14 January, with reports that beer in many pubs had frozen as temperatures fell as low as -24.6°C, and that there was a shortage of blood in many hospitals. Only once in the entire century had a marginally lower temperature been recorded. Many train, bus and air services were cancelled, diesel went "waxy" in the cold and would not flow at the pumps and with the sea icebound at Oban 'confused seagulls' could be seen 'to wander over frozen waves'. The savage cold saw the load on the South of Scotland's electricity supply exceed 4,400 megawatts, well above the previous peak in 1976-77.

In the midst of this misery, the hordes of striking workers were joined by a one-day stoppage by ambulancemen and the railway network then ground to a halt, not because of snow, but the threatened twenty-four hour walk-out by 28,000 train drivers. Lorry drivers were meanwhile still picketing and claims by Agriculture Minister John Silkin that eighty per cent of food supplies were getting through were greeted with derision by food manufacturers. A spokesman for Britain's biggest manufacturer of margarine and fats, Van den Bergh, said it 'hadn't made a delivery in a fortnight apart from small amounts to hospitals and schools'.

"The winter of discontent" was now really gathering pace. Eight million Londoners were again left without ambulance services for twenty-four hours on 19 January and a spokesman for the striking ambulancemen, Bill Dunn, was universally condemned for his comments on their refusal even to answer emergency calls: 'If it means lives are lost, that is how it must be. This time we are determined that the capital will take notice of what we are saying.' There was further cheerful news with the announcement that a Day of Action by 1.5 million public service workers planned for 22 January would affect burials, rubbish disposal, schools, hospital treatments and operations.

On 21 January snow returned, bringing widespread disruption once more. Dartmoor, Exmoor and Portland in Dorset were again cut off and air traffic at Heathrow disrupted. At midnight that night the snow-clearing crews at the airport joined other public service workers' in staging the "Day of Action" - a twenty-four hour strike. Hostile public reaction and pressure from Bill Dunn's own colleagues saw the ambulancemen's ban on emergency calls lifted but other ambulance services remained affected and every public service from airports to rubbish tips was shut down. The Times would have been expected to produce a ferocious editorial ... except that it was itself in the midst of a twelve month "lock-out" that heralded the beginning of the end for the old Fleet Street and a move by News International to a non-union plant at Wapping. Talks to end the lorry drivers dispute broke down and after

*Two kids making first use of a fresh fall of snow in a city park.*

*The start of one of the most notorious stretches of road in bad weather - the Stainmore Pass linking Durham and Westmorland (now Cumbria)*

a one-day strike on 23 January by train drivers, their union leader, Ray Buckton, walked out of talks with the British Railways Board and announced another twenty-four hour stoppage two days later. The ugliness of the industrial situation was shown by the news that sixty-five patients from a Birmingham cancer ward were sent home because of action by pickets of the National Union of Public Employees, and a hospital official claimed that a patient needing 'urgent treatment for a spinal disease' had been turned away from the hospital by pickets. Surgeon Mr Patrick Chesterman staged his own one-man protest at his outpatients clinic at Battle Hospital, Reading, by turning away anyone who was a union member. 'I decided it was time someone in this country hit back at the unions', he said, 'instead of taking everything lying down.'

Government figures showed that unemployment had now increased by 91,000 in a single month, with the blame laid equally on the strikes and the snow. There was plenty more of both to come. A fresh blizzard swept the country on 23 January, dumping four inches of snow on London, seven inches on the West Country and a foot of snow on parts of Wales. The search for an aircraft missing after crashing into the sea five miles off Dungeness on the Kent coast was abandoned as hopes faded for the lives of the five people aboard. Heathrow, Gatwick and half a dozen other airports were closed, and conditions on the roads were described as 'horrific', 'incredibly dangerous', atrocious' and 'diabolical', with abandoned cars 'littering' roads. As cars ploughed into each other in 'a chaotic spate' of accidents, thirty mile-an-hour speed limits were imposed across most of the motorway network. Nonetheless police patrols reported that many drivers were travelling at speeds of seventy miles an hour or more.

There were twenty-mile tailbacks on the M1 and travelling into London by car became so difficult that employees who rang their work were told 'Don't worry, turn around and go home'. Thousands did so, 'including jurors and witnesses at trials which had to be delayed or abandoned'. Branches of the major banks and some stores closed because of staff shortages, many schools were shut and many areas were without phone-lines after the weight of snow brought poles and lines down. In total it was estimated that about half of the eight million office, shop and service industry employees in the South-East took the day off and the pattern was repeated throughout the country as the rail strike and snow turned Britain into what the RAC termed 'a white hell'.

The one-day strike by train drivers made the effects of the weather even worse, and British Rail warned that the legacy would be felt the next day as well, particularly for commuters. 'Because no trains have been running today, heavy ice deposits have built up on conductor rails. While every effort will be made to get the lines open, our advice is not to rely on commuter services.' The one piece of cheering news was that a Boeing 707 carrying luxury goods and cases of whisky for the notorious Ugandan dictator Idi Amin and his entourage, had hit a patch of ice and skidded off the runway at Stansted when making the biweekly "whisky run" to Kampala. The Ugandan Airlines plane was buried up to its engines in snow ploughed up by the skid.

The forecast chaos on the railways, particularly British Rail's Southern Region, duly arrived the following day with hundreds of thousands of commuters stranded when the majority of services into London were cancelled because no de-icing had been possible on frozen points and conductor rails because of the previous day's strike. The few trains that ran were so full that many passengers were left on the platforms. At East Croydon, 'the crowd on the platform was fourteen-deep.' Most commuters still made efforts to get to work. Victoria Underground station was so overcrowded that it was closed to ease the pressure, buses were 'packed' and journeys that normally took thirty minutes were lasting two and a half hours. Sixteen year-old Shirley Billis showed extraordinary determination not to miss her day's work as a Fleet Street secretary. Leaving her home in Abbey Wood at 7.15am, she arrived at her office 'five and a quarter hours later and £7 poorer' having taken a taxi from Woolwich to Charing Cross. The railways were restored to some sort of order by the end of the day, just in time for the start of the next twenty-four hour strike by train drivers, so nobody got to work by train at all that day.

Council gritters were on strike in Southwark where 'ambulances were busy picking up people, mostly elderly, who had been injured by falling on icy pavements.' In Liverpool, health

in Loughborough after a bus slid off a snow-covered road and hit the queue of people waiting to board.

The fresh storms led Hannah Hauxwell, owner of Low Birk Hatt farm in Baldersdale, County Durham and famous as the eccentric star of the Yorkshire Television documentary "Too Long A Winter", to remark that 'This is the worst winter I have ever experienced. Last week was the roughest. On the Tuesday the electricity went off and it didn't come on again until Thursday. The wind got up and created a sort of whirlwind with the snow. Round here we call it "stouring". There was a roaring round the house... that was a bit frightening. While the electricity was off, I stayed in bed as long as I could, wearing all my woolly jumpers, my socks and a tweed coat. I couldn't heat water for my hot-water bottles - I like my bottles - and I couldn't make tea. I made myself corned beef sandwiches and drank cold water.'

The owners of the Tan Hill Inn, "Britain's highest inn" a few miles to the south, on the Pennine border between Cumbria, Durham and North Yorkshire, were once more cut off, and in total were 'snowed in for all but five days between New Year's Eve and April Fool's Day'. Snowdrifts buried the building up to the first floor windows and in between the blizzards, freezing rain covered every exposed surface in a two-inch layer of ice. While trying to move a generator that was frozen to the ground, the landlord slipped and broke his front teeth on the casing.

'For a week, I drank soup and bottled Guinness through a straw, yelling blue murder every time cold air hit my exposed teeth. When the weather at last cleared, I set off to walk the five miles to the edge of the moor - the road had been dug out to there by snow-blowers and JCBs - dragging a sledge behind me to fetch supplies. Some of the drifts must have been well over twenty feet deep, completely filling stream-beds and gullies, and when I reached the lane at the edge of the moor I was walking level with the tops of the walls because the road was completely buried under the drifts. A kindly farmer gave me a lift to a dentist in Barnard Castle, and having got a temporary repair to my teeth and bought as many groceries as I could fit on my sledge, I towed it back over the snow to Tan Hill, arriving exhausted just as the sun was going down.'

After being blocked in for six weeks, a snow-blower finally broke through to them. They jumped in the car and followed him back down the road to Reeth in Swaledale, 'bouncing between the banks of hard-packed snow as if we were tobogganing on the Cresta Run. We filled the boot with groceries and set off back to the inn. The wind was already blowing the snow back into the road as we drove up there and by nightfall we were once more cut off.'

On 15 February, the Callaghan government, accused of paralysis in the face of the country's industrial strife and desperate to be seen to be doing something about the weather, if

*'This is the worst winter I have ever experienced.'*
*Hannah Hauxwell, star of the Yorkshire Television documentary "Too Long A Winter"*

nothing else, now called in the avuncular figure of Denis Howell. Drafted in as "Minister for Drought" during the prolonged dry summer of 1976, Howell was now reincarnated as "Minister for Snow" and told to 'drop all other duties and co-ordinate efforts to keep the nation moving'. He was reported to be considering calling in troops to grit the roads and reminded councils of their 'obligations' to keep roads open, saying that help was available if needed. However, he turned a deaf ear to appeals to declare a State of Emergency and when asked what his immediate plans were said 'Well, I've offered my prayers to the Almighty - we'll have to wait and see if there's a response.' There almost was, as the helicopter carrying him on a visit to 'badly-hit Sheffield', almost collided with a private plane over Hertfordshire. There was more grim news for government and public with coal and electricity prices set to rise by nine per cent and inflation reaching 9.3 per cent, fuelled by the soaring price of foodstuffs because of the bad weather, but there was one straw to clutch at, with the suggestion that the long-running "dirty jobs" dispute with council manual workers was 'likely to be over' by Monday 19 February.

Peterborough Council did not wait for that, and that weekend called in private contractors to grit the roads after their twenty-five striking drivers had demanded £50,000 - albeit paid to charity - to suspend their own industrial action to do so. After the private contractors were brought in, the National Union of Public Employees announced that the council's entire 400-strong manual workforce would strike in retaliation. The news that a quarter of a million civil servants would be walking out in a separate pay dispute later that week caused rather less alarm to the public and led some comedians to speculate whether anyone would notice.

The AA meanwhile reported that dozens of main roads in Cambridgeshire, East Anglia, the Midlands and the North were still blocked by snow, and in Scotland a combined police and AA operation rescued sixty people, including twelve children, trapped in their cars by thirteen-foot snowdrifts near Dalwhinnie in the Grampians. With visibility in the blinding snow down to seven feet, the rescuers had to abandon their vehicles on the A9 Stirling to Inverness road and struggle through the snow for three miles in the gathering darkness, before 'probing through the snow with steel rods by torchlight' to find the abandoned vehicles. Some drivers, especially those with young children, opted to stay in sleeping bags in their cars rather than struggle in white-out visibility to Dalwhinnie. Once more the newspapers could not help pointing the contrast between Britain's chaos and the situation in Moscow, where despite temperatures of -30°C, 'everything was running like clockwork' according to a Thomson Holidays rep looking after 260 British holidaymakers. Meanwhile in China, the Peking Daily had more serious problems to deal with, after identifying 'yellow bell-bottom trousers' - the Chinese preferred to call them 'trumpet trousers' - 'long hair for boys' and 'fancy knitted caps' as a new threat to the morals of Chinese youth. 'How can this sort of thing help modernisation and socialist construction?' a correspondent of the People's Daily asked.

Back in Britain, on 23 February troops broke though to Cam Houses Farm near Hawes in Wensleydale, where farmer Robert Middleton and his family had been cut off since Christmas. He had broken his leg in December and had it put in plaster, but the snow had prevented him from going to hospital to have the plaster removed. His wife Susan related that he had telephoned his doctor, who told him he could take the plaster off himself, 'so he cut his leg out with a hacksaw'. Once a way had been opened to the farm, Mrs Middleton set off for the shops in the nearest town of Hawes, accompanied by her daughter Helen, 4, and

*"The Winter of Discontent" - Pedestrians are forced to pick their way over mounds of stinking rubbish dumped in a Glaswegian alleyway.*

riding on an army Scorpion tank. Men of the Royal Scots Dragoon Guards 'who took "Amazing Grace" to the top of the hit parade' - worked through the night to deliver six tons of fodder for the Middleton's starving sheep, and they left behind a souvenir, another Scorpion tank that became stuck in a huge snowdrift near the farm and 'will probably have to stay there until the snow melts'. At the hamlet of Booze in Swaledale a few miles to the north, troops used two twenty-ton diggers to cut a way through to three similarly isolated farms and take food to the farmers and fodder for their stock.

Like the weather, the industrial situation showed little sign of a thaw. With election day only a few weeks away, Prime Minister James Callaghan offered the National Coal Board a government subsidy to help them meet the cost of a settlement of the miners' forty per cent pay claim, but the next dispute was never far away. On the same day teachers' unions decided on a pay claim of 36.5 per cent, and when manual workers at three London hospitals went back after a twelve day strike, they promptly walked out again because the work areas they were returning to had not been properly cleaned. An IRA "bomb-blitz' on Yeovil on 23 February merely added to the air of crisis, and fear of the "Yorkshire Ripper", still prowling the streets despite a massive police investigation and thousands of calls from the public with information, was an added stress on women, particularly in the North of England.

On 5 March there were the first signs of spring, with the temperature in London rising to 12°C and day-trippers packing the beaches at Eastbourne, Brighton and Hastings, but the Northern winter continued, with hail and snow in Scotland and floods in Yorkshire. When the River Ouse burst its banks at York, 180 guests at the Victory Hotel had to be hastily evacuated.

On Saturday 17 March, with British Summer Time beginning that night, heavy snow once again began falling throughout Britain, with blizzard conditions as far south as Sussex. Many new-born lambs were killed. 'They never even had a chance to stand up,' one Cumbrian farmer ruefully remarked. The news that the fox-hunting season had been 'the worst for years' was a less universal cause for regret. Anglesey was without power for thirty hours after power lines collapsed under the weight of snow, and in Wiltshire lightning blew a crater in the grounds of Marlborough College and 'blasted open a sewer, hurling a manhole cover several feet into the air'. In North Wales, Shon Devey, a twenty-six year-old, climber, spent the night on Cader Idris near Dolgellau. He not only survived but, to outward appearances at least, also disproved the local legend that 'anyone who spends the night on the mountain becomes mad or a poet'.

The North and North-East were particularly badly hit with all roads into Newcastle blocked for several hours. Drifts in places were twenty feet deep and around 1,500 cars and several hundred lorries were abandoned. In Consett, County Durham, a tea-break nearly turned into a disaster when the cabin in which six workmen at the British Steel plant were taking their break collapsed under the weight of snow. Colleagues dug them out and all were taken to hospital. Astonishingly, one of two men from who died in the blizzards had gone out 'without a coat and wearing only light shoes'. James Mitchell, 42, from Todmorden in Yorkshire collapsed and died in the snow only yards from the caravan where he lived. The other victim, Alfred Cross, also forty-two and 'a keen walker', was found dead at the side of the Snake Pass on the A57 between Sheffield and Manchester. Twenty motorists may not have been too distraught to be snowed in the Jolly Sailor pub at Moorsholme, North Yorkshire, for the entire weekend. One of them, a lorry driver, John Clarkson, cheerfully reported that they were 'short of bread and milk, but not beer'. He and his companions may have been even more pleased that the snowploughs did not reach the pub until too late for them to get to work on Monday morning, but motorists in Cleveland were less happy to discover, when returning to the cars they had abandoned during the blizzard, that others had been there first. A police spokesman revealed that one car had all four wheels removed and another's headlights had been pinched.

The weather turned even worse the next day in the North, with police telling motorists not to drive into blizzard-hit Northumberland and Cumbria, and there was 'virtually no access

*Even where the snowploughs managed to get through, abandoned vehicles caused ongoing problems.*

to Scotland' at all. Police advised motorists and particularly lorry-drivers to stay in Newcastle for the night, rather than try to drive North, because 'they just won't get through'.

By 26 March the traditional North-South divide was once more in evidence with South-East England basking in temperatures of 15°C, the warmest day of the year so far, while torrential rain, and blizzards of sleet and snow were still sweeping the North. It was the last flurry of the coldest winter since 1963, but 1979 - "The Winter of Discontent" - had one final sting in its tail. In the General Election on 4 May 1979 Margaret Thatcher swept to power on a platform that included a pledge to curb the power of the unions. Neither winters, nor the political landscape of Britain have ever been the same since.

## EPILOGUE

# Could It Happen Again?

Over the past few years we've been enjoying a very mild run of winters. The prevailing wind has been westerly or south-westerly, and autumns have been warmer and lasted longer, winters shorter and milder, with Spring beginning earlier and earlier. This is because our weather has been dominated by the Atlantic, artificially warmed by the Gulf Stream, ensuring our shores remain mild. It's very different from the winters of 1947, 1962-63 and 1978-79, when intensely cold high pressure areas formed over Russia and Scandinavia. Cold air is much denser than warm air and these Russian "highs" tended to persist all winter, acting as a block to Atlantic weather systems moving eastwards, hence the term used by weathermen: "blocking high".

As a result our weather was dominated by intensely cold east or north-easterly winds, and as Atlantic weather systems attempted to push into the UK from the southwest, the rain associated with them readily turned to snow, with strong winds causing severe drifting, before the Russian high pressure zone pushed the Atlantic air away southwards, reasserting its icy grip. This happened repeatedly, giving the typical pattern of alternating periods of heavy snowfall and dry, cold weather with severe frosts and biting easterly winds.

The current run of very mild winters may be due to the fact that, as a direct result of global warming, Atlantic weather systems have more energy associated with them and travel further east, thus not giving the Russian highs the chance to dominate our weather as they did in the past.

In 2005, the Met Office predicted a colder than average winter based on a forecast of pressure anomalies between the Azores and Iceland, called the North Atlantic Oscillation (NAO). In fact, although much of Europe did have a colder than average winter, for much of time the UK was on the periphery of the very cold air in Europe. We had an early taste of snow when the main A30 through Cornwall ground to a halt after an unusually heavy snowfall for early winter, much to the excitement of the locals at the famous Jamaica Inn who stated that it was the worst early snow since the infamous winter of 1962/1963. However, not for the first time in the last few years, the early promise of a hard winter proved false.

The science of climate change seems to suggest that cold, snowy winters may be a thing of the past, although there are dissenting voices, and our climate has undergone cyclical temperature changes over thousands of years. The Vikings colonised Greenland during a warm period in the earth's history and grape vines were common as far north as

*A majestic winter scene, but the prolonged snowfall caused avalanches that claimed lives.*

*Ice floes drifting downriver as spring at last arrives.*

Knaresborough in Yorkshire. However our climate has also been much colder in the past, and the 1700s and 1800s were famous for their harsh winters when rivers across the country regularly froze. As recently as the 1960s and 1970s, the hard winters and dismally cold and wet summers led some scientists to believe that we were slipping into another cold period.

That now seems aeons ago, and the 1990s and the early years of the 21st century have seen a warming trend that cannot be explained by natural phenomena alone. The ten warmest ever years globally have occurred since 1990. Mean temperatures are rising, as is the concentration of carbon dioxide (CO2) - known to cause warming of the earth's atmosphere by allowing heat from the sun to warm the planet, but preventing it from re-radiating back into space. Yet without some C02, life would struggle to exist on earth as temperatures would be around 20C lower. The concentration of C02 in our atmosphere has recently been recorded at 380 parts per million at the U.S. observatory in Hawaii. This compares with about 200 ppm before the industrial revolution and, at least in part, is due to man's burning of fossil fuels such as oil and coal. Scientists have also recently discovered that the concentration of methane in the atmosphere has been underestimated and may account for an effective equivalent of 40ppm CO2. This would give a worldwide effective concentration of 420ppm CO2.... and some scientists believe that 400ppm is the point at which climate change becomes irreversible.

The rising levels of CO2 are only part of the problem; water vapour is by far the most potent of "greenhouse gases". The warming of the atmosphere causes an increase in the evaporation rate of our oceans, streams and rivers, releasing yet more water vapour into the air. This in turn causes more warming and the cycle starts to feed back on itself, eventually reaching a point of no return - "the runaway greenhouse effect" where the climate will change irreversibly.

So what does climate change really mean for winter weather patterns here in the UK? Mainstream thinking suggests that autumn and winters are likely to become milder and wetter. As the earth warms up it naturally follows that the sea will become warmer which will act to keep our country milder than normal since we are an island and surrounded completely by water. Weather systems may also become more vigorous, aided in part by the rising temperature of the sea, with stronger winds producing more frequent and violent storms. Episodes like the Burn's Day Storm of 1990, when large parts of the North of the UK, and in particular Scotland, were battered by winds topping 100mph, may become more frequent. There will be periods of heavier rainfall, leading to more flooding, like that of Autumn 2000, across a large part of the UK.

Climate change is also likely to mean big swings from one extreme to the other. Deeper areas of low pressure will generate stronger winds and heavier rainfall, but there has to be an equal and opposite reaction to this, because in a closed pressure system such as that of the earth, pressures must balance out. So there will also be intense areas of high pressure. These tend to be slower moving, leading to longer periods of dry weather, and the possibility of more drought. This has been evident in 2006, as anticyclonic weather has dominated weather patterns across southern parts of the country, leading to the threat of stand-pipes on the streets, a sight not seen since the drought of 1976.

However, a small but growing minority of scientists believe that global warming may actually lead to very much colder winters. Our shores are kept artificially warm by the action of the Gulf Stream, a huge conveyor of warm water stretching from its source in the Gulf of

Mexico to our coastline. If it wasn't for this, Britain - on the same latitude as Canada - would have a very much colder climate and because we are surrounded by water, it would be pretty grey and miserable as well, with copious amounts of snow.

The new theory points out that the strength and direction of this Gulf Stream is related to the salinity, and hence the density of the sea water. But global warming is causing huge quantities of fresh water to pour from the melting ice sheets of the Arctic Circle. Fresh water dilutes the seawater, causing its density to decrease. In time this could lead to an abrupt shutting down of the Gulf Stream somewhere in mid-Atlantic. Computer simulations suggest that the UK's annual temperature would then cool by up to 5 degrees Celsius in a matter of a decade or two. But the effect on extremes of temperature would be worse. Daily minimum temperatures in Central England could regularly fall well below minus 10C. If this were to happen disruption to society would be enormous, with agriculture, transport and other infrastructure severely hit.

How likely is this to happen? A recent report in the New Scientist stated that the ocean current that gives Western Europe its relatively balmy climate is already stuttering, raising fears that it might fail and plunge the continent into a mini ice age. The dramatic finding comes from a study of ocean circulation in the North Atlantic, which found a steep reduction in the warm currents that carry water north from the Gulf Stream.

At around 40° north – the latitude of Portugal and New York – the Gulf Stream current divides. Some water heads southwards, while the rest continues north, leading to warming winds that raise European temperatures by 5°C to 10°C.But when a team from the National Oceanography Centre in Southampton measured north-south heat flow last year, using a set of instruments strung across the Atlantic from the Canary Islands to the Bahamas, they found that the division of the waters appeared to have changed since previous surveys in 1957, 1981 and 1992. They calculate that the quantity of warm water flowing north had fallen by around 30%. When they added previously unanalysed data – collected in the same region by the US government's National Oceanic and Atmospheric Administration – they found a similar pattern. This suggests that his 2004 measurements are not a one-off, and that most of the slow-down happened between 1992 and 1998.

Harry Bryden from the National Oceanography Centre, speculates that the warming may have been part of a global temperature increase brought about by man-made greenhouse gases, and that this is now being counteracted by a decrease in the northward flow of warm water. After warming Europe, this flow comes to a halt in the waters off Greenland, sinks to the ocean floor and returns south. The water arriving from the south is already more saline and so more dense than the Arctic seas, and is made more so as ice forms.

Bryden's study has revealed that while one area of sinking water, on the Canadian side of Greenland, still seems to be functioning as normal, a second area on the European side has partially shut down and is sending only half as much deep water south as before. The two southward flows can be distinguished because they travel at different depths.

Nobody is clear on what has gone wrong. Suggestions include the melting of sea-ice or an increased flow from Siberian rivers into the Arctic. Both would load fresh water into the ocean surface, making it less dense and so preventing it from sinking, which in turn would slow the flow of tropical water from the south. And either could be triggered by man-made climate change. Some climate models predict that global warming could lead to such a shutdown later this century.

*Fresh snow and rising floodwaters at Marbury Mere in Cheshire.*

The last shutdown, which prompted a temperature drop of 5°C to 10°C in Western Europe, was probably at the end of the last ice age, 12,000 years ago. This is thought to have been due to fresh meltwater from a huge glacier in Canada, which flowed into the North Atlantic and stopped the ocean's sinking mechanism. There may also have been a slowing of Atlantic circulation during the Little Ice Age, which lasted sporadically from 1300 to about 1850 and created temperatures low enough to freeze the River Thames in London.

Harry Bryden says he is not yet sure if the change is temporary or signals a long-term trend. 'We don't want to say the circulation will shut down,' he told New Scientist. 'But we are nervous about our findings. They have come as quite a surprise.'

## Acknowledgements

Many individuals and organisations helped us in the preparation of this book. At Great Northern Books, our grateful thanks to Barry Cox, Patricia Lennon, David Joy and David Burrill; to David Barnett, Features Editor at the Telegraph & Argus, Bradford; and to Mike Baker; David Bailey-Bristol; Malcolm Beaton; Rod Bowen; Shaun Carroll; Hugh Conway-Jones; John Cundall; Marian and Peter Down; Aneurin Evans; David Fisher; Alan George; David Gott; Geoff Kirby; Geoff Matsell; Ian Mean; David Nicholson; Ed Perkins; Helen Reynolds; Robert Thomas; Ken Tytherleigh; David Wakeley; and Anne Wood.

## Photographic Credits

Every effort has been made to contact owners of copyright material. The following photographers and agencies are gratefully acknowledged:

Alamy Images; Associated Press; Jim Bray; Fox Photos; Getty Images; Kemsley Picture Service; Keystone Press Agency; Northern Echo; PA/Reuters; Planet News Ltd; A J Ransome; Topical Press Agency; West Riding News Service

# Subscribers

M G Longan
M G Longman
Martin Lonsdale
Elizabeth Lougheed
Pamela Lyons
Robert Lythe

John W Machin
Bill Mackintosh
J E Mallinson
Mandy & Moly
Graham Manifold
R L Mansfield
Chris Marks
Kenneth Marriott
Delia Marsden
Mr Eoric Marsden
Mrs Shirley Mason
Mr John Maud
Brian C Maw
G F Maw
Shirley Mellish
Mr A P Millard
Adam Milner
Frank Milner
R M Mitchell
Nicholas Monks
Jean Moody
John Moody
Dorothy Moore
S M Moore
Geoffrey Moorhouse
Mrs Anne Moscrop
Mr Paul Murgatroyd

Doreen Needham
Peter Newall
Vera Newhouse
M Newton
R Nicholls

E A Nickerson
Bryn D Niesyty
Josh Nixon
H Noble
Mary Norburn
Mr Alec North

Aena E O' Reilly
H M Osgerby
J Osgerby

Christopher J Page
Des Palmer
F L Parker
Sylvia Parsons
Raymond Pattison
Margaret Payton
John Pearson
Lovella Pearson
William Pearson
Douglas Philip Peck
David Peel
Dr Mark Pegden
Mrs Doreen Perkins
Mr G J Perry
David Pick
Jean Pickard
John D Pickard
Nora Pickard
David Piercy
Anne Poole
David Porter
Mr & Mrs David & Frances
Powl
Kenneth Price
Paul Price
Cyril Pridmore
Margaret Proctor
Timothy Prosser

M T Quirk

Alan Rae
Gwen Rawling
Sheila Rawnsley
Gordon Reynolds
Carole Rhoades
Shelagh M Richards
Linda Richardson
Gerald L Ricketts
Elizabeth Ridley Thomas
Hazel Rishworth
Denise Avis Ritchie
Elisabeth Roberts
Joan Roberts
Dr G Robinson
Frank Matthew Blackburn
Robinson
Philip Robinson
Stan Robinson
Gordon & Pruno Robson
Molly Rock
Paul Roddison
June Roebuck
Pamela Rollin
Margeret Ronson
J & N Rowland

Erika Salmon
Mary Sanderson
S P Saville
Patricia Schofield
Jack W Scott
Mrs C D Seipp
N S Sellers
Eileen Senior
Shelagh Senior
Mrs M Sharp
Barbara Sharpe
Derek Sharpe

Keith Shaw
Sadie Shaw
Robert Sheldon
Bill Shepard
Mary Sidgwick
Alan Simmester
Eric P Simpson
Jacqueline Simpson
Joan Simpson
Mr Tony Slater
Eric Sleight
William J Sleight
Walter H Slingsby
Christine Smith
Derrick Smith
Harry Smith
J B & B Smith
Joan & Bill Smith
John H Smith
Mr J R Smith
Brian Snaith
Mr Collin Snowden
Graham & Peta Snowdon
Jill Spensley
Steve Sprig
Charles Starkey
Mr C Starkey
Mrs Jean Mary Stear
Mr R H Steel
Jill Steele
F R A Stirk
M D Stoney
Barbara Stott
Chas Stuart
David Sturdy
Liz Summers
Theo Sumners
Andrew Swallow

Arthur Swallow
J A Swann
Peter Sykes

Denise Tabor
Mark Tabor
Mr Leslie Talbot
Thelma Tate
D H Tattersall
J D Taylor
Marjorie Taylor
Roger Taylor
John Terry
G S Thackray
Sally Theaker
Geoff Thompson
Graham Thompson
Max Thompson
Joan M Thomson
Stephen Thornton
Douglas Tovey
Nick Tovey
Maureen Treacy
David D Turner
Mrs Marie Turner

Bill Veitch

J E Waddilove
Mrs Kathy Wainwright
A V Walker
Les Walker
Marie Walker
David Wallace
Neil A Wallace
J Walmsley
Brian Walshaw
Anne Walton

Ian J Ward (Thorner)
Steve L Ware
Betty Watson
Cynthia Watts
Brian Weatherhead
Roger Weatherill
Mary Welbourn
Colin West
K West
D Westmoreland
Margaret Wharton
Betty White
The White Family
Joan Whitelam
Peter Whiteoak
David W Whitfield
Donald Whitfield
Valerie Whitfield
Brian & Christine Whittle
D J Whyley
Eric Wild
Marjorie Wilks
Mr W E Willey
Freda Williams
E Wilson
Paul Winterburn
Sue Withyman
Mrs E Wood
Phyllis Wood
Mrs Margaret Woodings
Dorothy Worrall
G D Wright
Mrs G Wright
Ernest Wriglesworth

Gerry Yates
Mrs V H Youell